SW

THE STORY OF INVENTIONS

FROM ANTIQUITY TO THE PRESENT

© 2008 h.f.ullmann publishing GmbH

Editor: Ritu Malhotra
Design: Mallika Das
DTP: Neeraj Aggarwal

Cover: All images: © Getty Images
Cover design: Simone Sticker
Overall responsibility for production: h.f.ullmann publishing GmbH, Potsdam, Germany

Printed in China, 2013

ISBN 978-3-8480-0638-0

10 9 8 7 6 5 4 3 2 1
X IX VIII VII VI V IV III II I

www.ullmann-publishing.com
newsletter@ullmann-publishing.com

Shobhit Mahajan

THE STORY OF INVENTIONS
FROM ANTIQUITY TO THE PRESENT

Contents

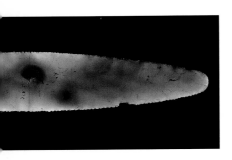

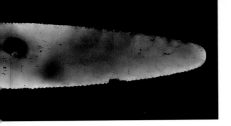

PREHISTORIC TIMES

EVOLUTION OF HUMANS

It is believed that the earliest human beings, or Homo sapiens, evolved about 200,000 years ago somewhere in Africa. These belonged to the human tribe Hominini, which had developed from some ape-like species in the Pliocene Epoch, 5.3 to 1.8 million years ago. There is no general agreement on which ape-like species gave rise to the early hominins but the fact that humans are related to apes is well-established and accepted.

The evolution of ape-like species to hominins was accelerated by the enormous climatic changes during the preceding Miocene Epoch, 11 to 5.3 million years ago. The latter part of this epoch saw a period of glaciation which led to major disturbances in the ecology of the earth. These disturbances included changes in the habitats by advancing glaciers—shifting of ocean currents and drying of vast plains. The evergreen forests were replaced by open spaces with shrubs and grass, and this provided the impetus for many evolutionary adaptations in the flora and fauna of the time. The shift from the dense forests to the drier savanna was responsible for major evolutionary adaptations in hominins. The most significant of these was bipedalism, or an upright stance. Bipedalism essentially freed hands; free hands could be employed to grasp and make tools, among other things.

Tools

The earliest fossil evidence of tool-making is from around 2.6 million years ago. These were simple tools made from stone; so this period in history is called the Stone Age. This is not to say that tools were not made before this; they possibly were, but in the absence of evidence it is conjectured that they were made of wood, bone, leaves or even grass—materials that do not survive the vicissitudes of time. There is evidence in the form of chimpanzees and orangutans using stems, vines and even stones (to break open hard fruits or bones) as tools to get food.

The earliest tools—chipped flakes made of flint and other stones—date back to the Paleolithic period, a term used to describe the early Stone Age when humans made a living by hunting animals and gathering nuts and berries (about 2.5 to 0.2 million years ago). These early tools were possibly made by hammering a pebble of flint—or any suitable, easily available fine-grained rock—with another stone and detaching a series of flakes till a jagged cutting edge was achieved. Such sharp-

Ancient Egyptian flint tool

Pre-dynastic Egypt had a well-developed material culture. The burial sites included fine pottery with representational designs, stone vases and mace heads. Cosmetic palettes with elaborate designs were also in use. Flint was worked with extraordinary skills into making many implements. Ceremonial knives were made from flint, a custom which carried over to the dynastic times. This finely worked flint implement is an example of the highly developed skills in working with stone. (Egypt; Pre-dynastic; Ashmolean Museum, Oxford)

Periods in prehistory are defined by a geologic timescale. These periods are marked by changes in the environment, which in turn led to the diversification of the flora and fauna, and several evolutionary adaptations.

5.3 to 1.8 million years ago: Pliocene Epoch
This period was characterized by cool and dry climate, and large mammals. Australopithecines, the earliest hominins, developed in this period. Important inventions included rudimentary stone tools.

1.8 million to 11,500 years ago: Pleistocene Epoch
Referred to as the Great Ice Age, this period is recognized for glaciation or development of large ice-sheets. Many large mammals flourished and later became extinct in this epoch. The most significant development was the evolution of modern humans.

1.5 million years ago: Acheulean Stone Industry—hand axes made with stone in many areas of the world.

500,000 years ago: Use of fire.

200,000 years ago: Emergence of Homo sapiens.

50,000 years ago: Bone, antler tools. Microliths make their appearance in stone tools.

12,000 years ago: Pottery.

11,500 years ago: Beginning of Holocene Epoch
This epoch marks the beginning of an interglacial period. The withdrawal of the ice-sheets to their present positions and increase in precipitation led to the rise of human civilization.

9000 BC: Domestication of sheep.

9000 BC: Use of sun-dried bricks for making houses in Jericho.

8000 BC: First use of copper.

7000 BC: Beginnings of agriculture. Wheat, barley and pea domesticated in some places. By 7000 BC, there were farming communities in the Middle East, Greece, Anatolia, Crete as well as the western fringes of the Indus Valley. Agriculture also spread through south and central Europe.

7000 BC: Domestication of rice and millet in China.

6000 BC: Molded bricks used on the Anatolian plateau.

4500 BC: Beginning of the pre-dynastic period in Egypt.

Human ancestors using stone tools for hunting

4000 BC: First attempts at producing synthetic material—Egyptian faience.

4000 BC: Pottery kilns come into use, making it possible to produce fired pottery on a large scale.

4000 BC: First use of a seal—small circular discs of fired clay or stone with an impression.

dged tools, called choppers, were probably used by early hominins to cut meat, remove the hides from dead animals, and even shape sticks to be used for other purposes.

Around 1.5 million years ago, the most versatile tool of the early hominin, the hand axe, was invented. Though the hand axe might have made its first appearance anywhere, it is usually associated with a tradition of tool-making called Acheulean, after the site Saint-Acheul in northern France. The Acheulean period, which lasted from 1.5 million to 200,000 years ago, was characterized by the use of many different kinds of stones—flint, sandstone, chert, shale, basalt (depending on the area)—for making tools. The Acheulean industry was very widespread, with sites in Africa, Europe, the Middle East and even Asia.

Being a versatile tool, the hand axe was a popular invention that remained in use for a long period of time. Found throughout Africa and in large parts of western Asia as well as Europe, it may have been used for chopping, cutting or digging up roots. The earliest hand axes were crude, sharp-edged tools, made from a fine-grained stone by hitting it against another stone. With little variation in form, the tool gradually became more refined in terms of straighter or even saw-toothed edges.

During the later part of the Acheulean period, the early hominin started paying attention to the flakes which fell off during the making of the hand axe and other tools. These flakes could be used either directly as knives or with some minimum shaping as side-scrapers or other tools. At some point in time, these flakes were being set in handles or hafts to serve as spearheads or knife blades.

Gradually, more sophisticated ways of making stone tools were developed. For instance, narrow blades of flint were obtained from a block and shaped into specialized tools such as knives, scrapers, chisels and gravers. Other materials were also being used for making tools. During the late Paleolithic period—50,000 to 20,000 years ago—bone, wood, antler, ivory and shell are known to have been used for making tools. These tools were much more specialized and superior to the early tools used by hominins. There is evidence of the use of fish hooks as well as needles made from bones.

Even with stone, the tools were now more sophisticated with the use of microblades in spearheads and harpoons. The tools became composite, requiring a fabrication from diverse elements; examples include harpoons, bows and arrows. The presence of perforated bone needles indicates that some kind of weaving or sewing was also being practised during this period.

Acheulean hand axe

The first type of recognizable tool made by man was the hand axe of flint. It was a multipurpose tool used for scraping, cutting and chopping. Hand axes could be made from flint or any other kind of stone which could provide a sharp edge. Starting with a piece of stone, flakes were hammered off around the edge to produce the axe. Flakes were later used by themselves as they were shaped into knives and scrapers. (Tanzania; 60000 BC; Stone; British Institue of History and Archaeology, Dar-es-Salam)

Fire

Apart from tool-making, the other technology which set the early hominins apart from their primate cousins was the control of fire. Fire in the form of lightning, causing burning of dried forests and shrubs, would have been a common sight. But controlling fire was not possible till about 500,000 years ago. There is some evidence of fire control in South Africa about 1.5 million years ago but it is not conclusive. However, charcoal, burned seeds and charred bones from around 500,000 years ago indicate that hominins were making and maintaining fire, though it was not until about 3000 BC that reliable fire-making techniques like fire drills and flints were mastered by humans. Fire was immensely beneficial to the early humans in cooking, in providing warmth and light, and also in guarding against predators.

NEOLITHIC REVOLUTION

Around 11,500 years ago, the last of the major climatic changes happened, marking the beginning of the Holocene Epoch. During this period, the ice-sheets, which had covered most of the northern hemisphere, withdrew to the region in which they are found today. The withdrawal of the ice-sheets resulted in geologic changes too. The melting ice led to an increase in sea levels, causing temporary flooding of areas which were far from the sea. This is attested by the presence of marine fossils in areas like Quebec and Vermont. The Baltic Sea

Neolithic implements

Stone was the material most used for making tools during the early Neolithic period. These drawings indicate some of the kinds of implements used by the Neolithic man. Stone axes with polished stone—polishing was done by rubbing against a piece of sandstone—could have been used to clear forests to plant grain. Obsidian and flint might have been obtained by pressure-flaking with a piece of bone or wood. Bone and antler tools could have been made by cutting them with flint knives and polishing them with suitable stones.

was also formed by the recession of the ice-sheets. The increased precipitation toward the beginning of Holocene in the equatorial regions also changed the character of the great rivers. There was an enormous increase in the discharge of rivers like the Nile, Congo and Niger during this period.

The climatic changes also affected the flora and fauna of many regions. For instance, the former tundra regions became populated with pine, or the erstwhile arid areas witnessed a spurt of moist savanna vegetation. Many large mammalian species became extinct in this period. The reasons for extinction could have been climatic or cultural, or both. As the rise in the temperatures allowed humans to move further north in Europe and Asia, the cultural reasons for the disappearance of mega fauna could be overhunting by a growing population of hunter-gatherers in habitats which were too inhospitable hitherto.

Agriculture

Around 11,000 to 9,000 years ago, human beings domesticated plants. This single event, o a series of events in different parts of the world, i possibly the most significant one in the history o humankind. Agriculture changed the way huma beings organized themselves. This marks th beginning of the Neolithic period, or late Ston Age, which led to dramatic social changes.

Neolithic culture emerged in the Levant aroun 8500 BC. It was influenced by the Natufians o Palestine—referred to as "proto-Neolithic"— who were using sickles to harvest either wild o sown grain around this time. Emmer and einkor wheat, barley, peas, lentils and chick peas wer some of the plants being grown or harvested i the Middle East in this period. It is possible, o almost certain, that in different agro climati zones different plants were grown for food— pumpkins and gourds in Mexico around 7000 BC or water chestnuts in southeast Asia in 9000 BC As the transition from the hunting-gatherin lifestyle to a more settled life was gradual an widespread, it is difficult to point to a single regio where agriculture started.

The earliest sickles to reap the grain were made o wood and flint; sickles made of flint blade set in th jawbone of animals have also been found. Initiall simple digging sticks were used to plant seeds these were later adapted to some form of a hoe. Th grain was made more palatable by grinding it o mills made from two stones. This was similar to th mortars used by cave dwellers to make pigments which were used in cave paintings of the kind foun at Lascaux in France. It was prevalent to stor grain in pit silos and granaries for use during th lean seasons.

Domestication of Animals

By around 7000 BC, the settled communities especially in the Middle East, were beginning t diversify and make efficient use of the resource available from the land. Domestication of animal began around this time. On the basis of calibrate radiocarbon dating of animal remains, it is know that domesticated sheep existed in northern

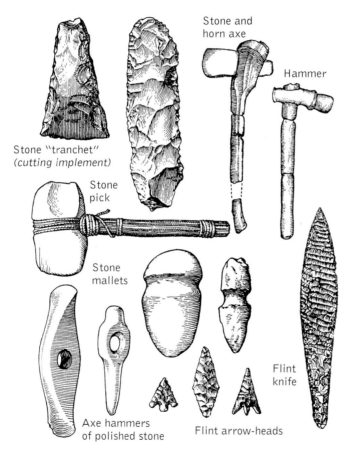

Stone "tranchet" (cutting implement)

Stone pick

Stone mallets

Stone and horn axe

Hammer

Flint knife

Axe hammers of polished stone

Flint arrow-heads

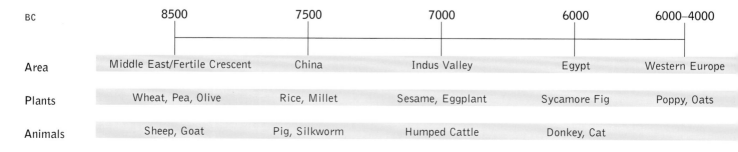

DOMESTICATION OF PLANT AND ANIMAL SPECIES

BC	8500	7500	7000	6000	6000–4000
Area	Middle East/Fertile Crescent	China	Indus Valley	Egypt	Western Europe
Plants	Wheat, Pea, Olive	Rice, Millet	Sesame, Eggplant	Sycamore Fig	Poppy, Oats
Animals	Sheep, Goat	Pig, Silkworm	Humped Cattle	Donkey, Cat	

Iraq from about 9000 BC while the first known instances of domestication of cattle are dated around 6000 BC.

Domesticated animals enhanced the sustaining capacity of land in many ways—by providing readily available meat, milk and fertilizer, and later by pulling plows to increase crop yield. Apart from this, these animals provided materials like leather which could be put to diverse uses. Large mammals also had a profound impact on the mobility of humans, initially directly on them and later with the help of wagons. Trade, for instance, was facilitated by the means of domestic animals. At a later date, animals like the horse, camel and ox would have a major impact on wars and conquests.

Housing and Lifestyle

As the hunter-gatherers adopted a more settled lifestyle, their dwellings, which were previously just caves or natural shelters, gave way to mud-brick houses. Made of alluvial mud, these bricks were modeled by hand and then dried in the sun. Some of the early habitations on the Anatolian Plateau were made from sun-dried bricks but these were not hand-modeled. The uniformity of the bricks suggests that some kind of a mold was used to produce them. In the river valleys and deltas, where reeds were common, reed huts, and possibly reed matting, might have been used for housing purposes.

The settled communities which came up with the advent of agriculture were usually self-contained, and yet there was interaction between different regions. For instance, at Knossos in Crete, there

are signs of habitation from around 5000 BC. These people were doing rudimentary farming and living in mud-brick dwellings much like the communities on the Anatolian Plateau. Clearly, they must have reached the island of Crete—which is around 124 miles (200 km) from the mainland—in some form of seaworthy rafts or vessels. Humans were probably using rudimentary boats or rafts to cross rivers and channels much before this, but what these vessels were made of and how they were made is not known.

The first known pottery vessels were produced in Japan around 10,000 years ago during the Jomon period, which lasted from 7500 BC to 250 BC. Before that, vessels made of wood and stone were used. The use of clay to shape vessels was a significant technological advancement. Clay was mixed with other materials like sand or organic matter to prevent shrinkage during drying; after drying in the sun, the pots were fired in the domestic hearth or a bonfire and later in pit kilns.

There is evidence of weaving of mats and baskets from very early times. It is not known if cloth was being made from fiber at this early stage but there are many sites where primitive spindle whorls have been found. These indicate that thread was being made at an early date, but there is no record of woven fabric.

By about 5000 BC, villages with agriculture and domesticated animals were common in the Tigris-Euphrates valley, Greece, Anatolia, Levant as well as Crete. In the next millennium, the technologies and practices of farming and other associated activities diffused to many different parts of the world and led to the flowering of the Ancient River Civilizations.

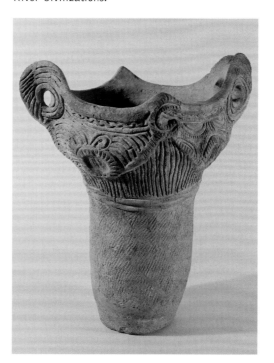

Female figurines

The emergence of pottery around the time that the Neolithic man started domesticating plants is possibly not a coincidence. Clay was used to make containers to store grain as well as to cook. Using the plastic qualities of clay to make figures, toy animals and models presumably started around the same time. These female figurines from the Moravia region in the eastern part of the Czech Republic are examples of the exquisite nature of some of the Neolithic art. (Moravia; Neolithic period; Baked clay; Moravian Museum, Brno)

Jomon vase

Lasting from 7500 BC till 250 BC, Jomon pottery from ancient Japan was a highly decorative form of art. Marked by regional and chronological variations, it reflected the culture of the people. The pottery of the early period, shown here, was typically in the form of deep vessels with tapering, bullet-shaped bottoms. (Kanto province, Japan; Earthenware; Musee Guimet, Paris)

FIRE AND FARMING

As a famous philosopher said, "Everybody has his own Prometheus"—so does every culture. Fire is as fundamental to all mythologies as cosmology. Several anthropologists feel that what distinguished humans from animals is that they exploited fire. However, the real origins of the domestication of fire are unknown. There are many theories about how and when this could have happened but none is conclusive.

Initially, fire was used for warmth, light and protection from wild animals. But it was the use of fire for cooking which was by far the most important innovation from a cultural perspective. It is possible that some form of cooking was practised by early humans long before fire was domesticated. Naturally occurring fires in forests could have produced roasted seeds and fruits which were readily foraged by many animals including chimpanzees. Early

A historical depiction of Stone Age baking over fire

humans might have observed and imitated this in the form of cooking.

The first use of fire was in all likelihood grilling and roasting of meat directly over the flames. This would have made the meat far more edible and digestible. Cooking over fire also destroyed various toxins present in some naturally-occurring food, making it fit for human consumption.

At some point in history, it was discovered that food could be cooked on hot stones, leading to a greater variety in taste. This form of cooking was particularly useful for food which comes wrapped in coatings which retain heat—fibrous husks,

some varieties of fruit and even mollusks. Later innovations in the form of cooking pits led to a rudimentary form of baking—by using a pit heated with stones—or a boiler (by using heated water).

Still later, utensils woven from fronds or grasses would have come into being. Hunters must have also used animal skin and tripe to cook over flames as is common in several cultures even today. With the invention of pottery, clay utensils for cooking became almost universal, and this was the case for a long time till metals began to be used. In fact, it is possible that the woven utensils were initially coated with clay to make them more durable and this gave the initial idea of using only clay (mixed with sand or gravel) to make cooking utensils.

Initially, fire wasn't readily available. It had to be collected by so-called fire gatherers who had to save it from rain and wind, and had to keep it going by constantly supplying fresh combustible material. Once it was accessible—possibly by striking two pieces of wood surrounded by combustible material—fire paved the way for technological advancements like smelting ore, firing pottery and creating complex metal weapons.

The other important invention that changed the way of life in prehistoric times was agriculture. Surviving as hunter-gatherers since the beginning, it was only in the last 11,000 years or so that people turned to food production and adopted a more settled lifestyle. Before this time, large, wild mammals were abundant, and thus hunting-gathering might have been easier and more efficient than growing and storing food. Of course, the main factor that facilitated this transition was the change in global environment. In addition, the wild cereals, from which the domesticated plants came about, were not yet available in large quantities. Also, the technologies for processing and storing food grains had not yet evolved.

Since the shift from hunting-gathering to food-producing lifestyle was gradual, it is likely that for some time both communities co-existed. As the nomadic lifestyle of hunter-gatherers led to a lower birth rate, the increased numbers of settled communities must have proved to be a military advantage in conflicts with hunter-gathering tribes. This and other advantages gradually made agriculture the dominant way of life.

Once humans started farming, the higher population densities meant increased demand for food and thus increased production. As food became surplus, it led to the rise of specialists, non food-producing sections in the society, such as bureaucrats and craftsmen (potters, brick-builders and so on). The organization of society into skill-based specialists led to the growth of a more complex economy and hierarchical society, and civilization as we know it.

ANCIENT CIVILIZATIONS

While it is true that agriculture demanded a much more settled life than the nomadic lifestyle of hunter-gatherers, the spread of farming communities indicates that early farmers migrated and influenced their nomadic neighbors to adopt their lifestyle. The result: farming communities spread within a few centuries from the Middle East to central Europe, the Indus Valley and down the Nile River.

The spread of farming communities along the river valleys was expected since these rivers not only served as channels of communication to facilitate the movement of ideas and materials but also made possible perennial cropping. The flooding of the rivers also brought with it silt which replenished the soil with essential nutrients, thus increasing its productivity.

The increasing productivity meant surplus food and increased population density, which led to the flourishing of other non-agricultural skills such as craft and trade. The rudimentary division of labor, over a period of time, made possible some major technological innovations. The increasing use of metal was among the most significant inventions at that time.

Use of Metal

Copper was the first metal to be used by humans. The evidence of its first use dates back to about 3000 BC. Copper, unlike iron, occasionally occurs in its metallic state in the form of lumps mixed with its ores. Copper ores are found in northern Syria and the neighboring Turkey, areas where farming communities were already established. So it is likely that human beings chanced upon metallic copper, attracted by its greenish color. The metal, unlike stone which humans were familiar with for thousands of years, could not be chipped to make tools. However, it could be easily bent and hammered, and was thus used for small trinkets initially.

Apart from copper, gold and silver were also in use, though very sparingly. Gold veins are found in rocks but are quite hard to extract. However, deposits of the metal are frequently washed down in rivers and could have been possibly found by people living near the rivers. Silver is even rarer

than gold and was usually found as a natural alloy with gold.

Apart from copper, gold and silver, one other material which was treasured around this time was lapis lazuli. This is found in very few places and it is possible that the vivid colors of this mineral made it attractive for decorative purposes. Nevertheless, as early as 4000 BC, there were attempts to recreate this in the form of what is now referred to as Egyptian faience. Making this possibly involved the heating of an ore of copper and a stone together to form a kind of glass. For this, the material had to be heated in some closed vessel and the fire temperature had to be much higher than usual. It is possible that the fire was blown or fanned, and this could have been done by the use of reeds.

The method of making faience paved the way for the most important innovation in technology around this time, that of smelting of copper from its ores. This one innovation had a tremendous impact on the development of societies in the next two thousand years. It not only led to the invention of bronze—a better material for making tools and ornaments—but also provided the technology for the invention of iron several centuries later.

Around 4000 BC, there is the first evidence of the use of pottery kilns in Mesopotamia. This allowed the separation of vessels from the fire and thus made possible more variety in the design and color of the pottery.

ANCIENT RIVER CIVILIZATIONS

As already seen, farming communities spread to many places from the Middle East after the beginning of agriculture. The spread along river valleys was significant since the natural conditions in these places were favorable for the flowering of the Great River Civilizations at a later date.

In ancient Mesopotamia, near the present Syria-Turkey border, Tall Halaf was a city which flourished from 5050 BC to about 4300 BC. Dryland farming was practised here, and there is evidence of wheat, barley and flax—the seeds of flax provide oil and its fiber can be used for making linen cloth—being grown. Their pottery is characterized by geometric and animal designs.

In northern Mesopotamia, near the present-day city of Mosul in Iraq, there is evidence of the first settled community at Hassuna. These people used hand axes, sickles, grinding stones and even baking ovens. Grain was stored in large clay vessels which were sunk in the ground. There are remains of jar burials in which food is placed, signifying some kind of belief in after life. Among the pottery found here is the Sammara ware which is characteristic of a region in Iran, thereby indicating that there were already extensive trade contacts or migration of workmen between the regions.

In southern Mesopotamia, Ubaid culture flourished in this period. The culture is characterized by painted pottery with floral and geometric designs. Some of the pottery seems to have been made on a slow-turning wheel rather than being hand-molded. The first instances of loop handles and spouts on pottery are from this region. The settlements are characterized by multi-roomed, mud-brick houses and temples in public spaces. In the latter part of 5000 BC, Ubaid artifacts were found all over Arabia, indicating a growth of a trading system.

The growth of farming communities was not restricted to the Fertile Crescent—the area extending from the Tigris-Euphrates River system to the Nile River—but extended beyond the Iranian plateau into the Indus valley. Mehrgarh is one of the earliest settlements in this region (8000–3000 BC). In the earliest phases, this was a farming community growing wheat and barley, living in mud-brick houses and burying their dead. Interestingly, even from these early times, there is evidence of stone-bead and shell necklaces being placed along with the human remains.

In the later period, there is evidence of the use of pottery and many structures for storage of grain, implying an increasing population and productivity. Ivory, copper and glazed faience beads were in use, and there is evidence of copper melting crucibles from this period.

The period before 4000 BC saw the settling down of humans in various river valleys as well as highlands. With increasing productivity of agriculture—due to irrigation, inundation of land by the flooding rivers and crop rotation—the settlements grew larger in size and hierarchical structures in society seemed to set in. The stage was thus set for the birth of the Great River Civilizations in the next millennium.

Carved slab from Nimrud

The earliest civilizations emerged in regions of Mesopotamia. These civilizations had an organized society and advanced lifestyle. This depiction of the camp of King Ashurnasirpal II shows people engaged in various domestic activities, reflecting the developed lifestyle in those times. (Assyria, Ancient Iraq; 865 BC; Stone; British Museum, London)

ANCIENT TO CLASSICAL

CIVILIZATION, SOCIETY AND CITIES (4000 BC–1000 AD)

ANCIENT MESOPOTAMIA AND EGYPT

The Neolithic Revolution led to the development from the Stone Age to the Bronze Age, marking the end of prehistoric times and the beginning of ancient times. Characterized by the use of bronze for making tools, weapons and ornaments, the Bronze Age lasted from around 4000 BC to the beginning of Iron Age in 1000 AD. It was a golden period of technological inventions: the wheel, the plow, writing, money and cities being a few important developments of the period.

The earliest of the ancient civilizations had emerged in the Mesopotamian region in the period preceding 4000 BC. Though Mesopotamia referred to the area between the two rivers, Euphrates and Tigris, it is generally taken to include eastern Syria, southeastern Turkey and all of Iraq. The hilly northern regions of this area—which receive more rainfall—were settled first, while the southern and central regions had a much more arid climate as well as hard and dry soil. In the southern regions, the two rivers usually overflowed because of heavy deposition of silt. However, the soil was extremely fertile, and by around 4000 BC, the southern region, with the help of canals and drainage systems, had overtaken the north in terms of prosperity. It became known for the civilizations of Sumer—famous for earliest writing called cuneiform—and later Babylon, where the first set of laws in recorded history, Hammurabi's Code, came into being.

Meanwhile, along the River Nile, civilization emerged in the valley of Egypt. Egyptians adopted many Mesopotamian innovations and developed many of their own—pyramids and a 365-day solar calendar being among the most significant ones. Supported by the River Nile and protected from invaders by the Sahara Desert, the Egyptian civilization lasted for 3,000 years (from 3100 BC to 30 BC), the longest in history.

Hieroglyphs on a tomb at Giza

A major instrument for centralized control in ancient Egypt was writing. Two kinds of writing were in use: hieroglyphs, which evolved from pictograms, were used mostly in monuments, and the hieratic, which was the cursive form. A part of the will of Kaiemnefert—an administrator of the fifth dynasty—on a stone block, is shown here. The most important use of writing was for administration, and before about 2600 BC, no continuous texts are known. (Giza, Egypt; Fifth dynasty/2494–2345 BC; Stone; Egyptian Museum, Cairo)

MESOPOTAMIA
4000 BC: Sumerians settle in Mesopotamia. City states established. First use of copper. Clay and stone seals introduced.

3900 BC: Ubaid culture.

3600 BC: The city of Uruk founded.

3500 BC: Sumerians invent the wheel. Copper casting developed. Cuneiform writing in Sumer.

3400 BC: First use of cylindrical seal in Sumer. Chariot invented.

2750 BC: First Sumerian dynasty of Ur.

2400 BC: Sumerians develop calendar.

2300 BC: Bronze-making begins.

2340–2125 BC: Sargon I begins the Akkadian rule and unifies the whole region into a single kingdom.

1800-1170 BC: Old Babylonian period. Code of Hammurabi written.

1600–1100 BC: Periods of Hittites hegemony. First use of iron.

1200–612 BC: Assyrian period.

612–539 BC: Neo-Babylonian period.

539 BC: Fall of Babylon and the beginning of Persian dominance.

EGYPT
3100–2950 BC: Late pre-dynastic period. Hieroglyphic writing and establishment of cities.

2950–2575 BC: Early dynastic period. First pyramid is built at Saqqara. Egyptians make leavened bread.

2575–2150 BC: Old kingdom. Great pyramids built at Dahshur and Giza.

1975–1640 BC: Middle kingdom. Upper and lower Egypt reunited by Mentuhotep.

1539–1075 BC: New kingdom. Peak of Egypt's rule.

332 BC–395 AD: Alexander occupies Egypt. After Macedonians and Ptolemies, Romans make it a province of Roman Empire.

GREECE
3000–1100 BC: Bronze Age. Minoan and Mycenean civilizations.

1100–800 BC: Dark age. Dorians form city states, including Athens and Sparta. Geometric schemes used on pottery.

800–500 BC: Archaic period. First Olympics held (776 BC). Homer writes the *Illiad* and *Odyssey*. Draco's code of law introduced in Athens. First use of coin currency. Persian invasion of Greek territories.

Ur III clay administrative tablet

500–330 BC: Classical period. Democracy in Athens. Greeks defeat Persian invaders and Athens becomes a powerful city. Famous Greek plays are written and the Parthenon is built. Sparta defeats Athens. Philip of Macedonia rules Greece; his son Alexander conquers the world.

330–30 BC: Hellenistic age. Greece becomes part of Roman Empire.

ROME
753 BC: Foundation of Rome.

509 BC: Rome becomes Republic.

450 BC: First law code—Twelve Tables.

378 BC: City wall built.

312 BC: First aqueduct and first major road built.

280 BC: Coinage begins.

200 BC: Use of concrete in Palestrina.

130 BC: Rome conquers Greece and most of Spain.

85 BC: Heating system introduced.

122 AD: Building of Hadrian's wall.

324 AD: Foundation of Constantinople.

455–476 AD: Vandals destroy Rome. Western Roman Empire falls to Visigoths and Ostrogoths.

554–1453 AD: Eastern Empire survives as Byzantine Empire.

ISLAM
600 AD: Spread of Islam.

661–750 AD: The Umayyad Caliphate.

750–935 AD: The Abbasid Caliphate. The founding of Baghdad.

Writing

Around 4000 BC, the first use of the seal is recorded from Mesopotamia. Early seals were made from fired clay or stone into which some geometric impressions were cut. The patterns on the earliest seals indicate that the device was used as an identification mark of the owner. The invention of the seal had two important implications—one, this set the stage for writing, a tool for recording things in a permanent way, and thus demarcating proto-history from prehistory. Secondly, the idea of producing a positive from negative—the final mark impressed by the seal on some wet clay which is an inverted image of the markings on the seal—became prevalent.

The use of seals as identification marks led to the idea of writing. Making marks on a tablet of wet clay, drying the clay and storing it led to a rudimentary form of writing. This is first seen in Mesopotamia around 3500 BC. The first markings on the tablets were pictogram representations of objects like sheep, indicating that writing originated from some practical necessity like accounting. Possibly styluses made from reeds were used to draw on the wet clay. The primitive pictograms evolved into more elaborate symbols as time went on. This writing system, known as

cuneiform, continued in use in Mesopotamia until around the first century AD.

Though clay was the material of choice in Mesopotamia, the Egyptians made use of the papyrus reed which grows in the Nile delta to make a paperlike material. The reeds were cut into strips which were laid crosswise and then beaten with a wooden mallet to make a thin material which could be rolled and thus stored easily. The pictograms on the papyrus rolls were made with ink, and these pictograms finally evolved into the hieroglyphs. Later, a cursive form of this writing called hieratic came into being.

Copper and Bronze

The period around 4000 BC also saw one of the most remarkable discoveries made by humans—extracting metallic copper from its ores. This one discovery had profound implications for the history of technology because it not only introduced a new, durable material for making implements and weapons but also made possible the use of other metals like iron several centuries later.

Reducing copper from its ore implied mixing the ore with some suitable fuel like charcoal or wood and placing this in a fire. Ordinary fires in the hearth or the pottery kilns couldn't have

Ancient civilizations

With the exception of the Roman and Greek civilizations and the civilizations in the Americas, all the others were essentially river valley based. The regular flooding of the rivers—the Nile, Huang Ho and Indus—provided the necessary nutrients to the soil to sustain the agricultural productivity required for feeding the urban centers.

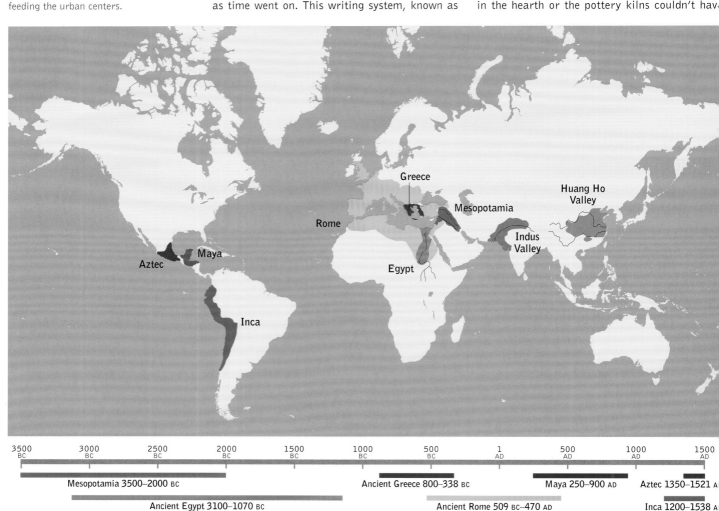

produced the temperatures required for smelting copper; hence it is possible that blow pipes (made of reeds) were used to stoke the fire and obtain the high temperatures needed. The molten metal would settle to the bottom while the impurities in the form of slag would need to be chipped away.

The metallic copper could have been hammered or broken up into smaller pieces and then fashioned into objects. Around 3500 BC, humans discovered an even better way of using copper. A piece of metallic copper placed in a crucible and heated in a furnace would melt, and this molten metal could be poured into molds (possibly made from stone) of the kind that had been used first in brick-making many centuries ago. The casting of copper made weapons more sophisticated, which hitherto had been made in a very crude fashion. Of course, there were many incremental improvements in the molding technology like using two pieces for the mold and so on. However, given the scarcity of copper ores in the areas around which the first settlements emerged, the use of copper was restricted to weaponry and that too by the wealthy people.

Around 3000 BC, possibly in the area near Syria and eastern Turkey, another new material was discovered. The artisans working with copper discovered that adding a little bit of another locally available material, namely tinstone, resulted in a much more useful alloy than copper itself. This discovery heralded the Bronze Age and led to an enormous improvement in the quality of weaponry and tools. Bronze is much harder and has a lower melting point than copper, thus making molding easier. Weapons, tools, utensils and other artifacts became more prevalent. Though Mesopotamia did not have any tin or copper, it did acquire the materials by trade as there is plenty of evidence of its use. About 500 years later, the lead ore Galena, which contains traces of silver, was smelted to obtain reasonable quantities of silver. The first true glass was possibly made in Mesopotamia around 2000 BC with the heating together of quartz and soda. The lime naturally present in the raw materials made the glass stable, and lead was added to the glass to give it more brilliance.

Housing and Monuments

It is likely that the experience from pottery kilns and copper furnaces led to the idea of fired bricks. In Mesopotamia, around 3500 BC, the fired bricks made urbanization possible on a huge scale. Temple complexes in Mesopotamian cities used a mixture of fired bricks, stone and sun-dried bricks, indicating that the use of fired bricks was still a novelty. However, the technological innovation did lead to a qualitative change in the building technology, though its use did not permeate down to ordinary people for several centuries.

The different styles of building in Egypt and in Mesopotamia were due to the nature of material available. The Mesopotamians used mostly fired or sun-dried bricks for their monuments while the lintels and the facing were made of stone, which had to be imported in this flat, dry land.

The Egyptians, in contrast, had access to stone quarries and hence the monuments were all built with stone. Sun-dried bricks were also used, for instance in the construction of burial chambers. The Egyptians started building pyramids in 3000 BC. What started off as burial chambers became more and more elaborate as time went on, and culminated in the Great Pyramid of Khufu (Cheops) at Giza, constructed around 2560 BC.

Mathematics and Science

The pyramids in Egypt and the ziggurats in Mesopotamia were magnificent architectural achievements. Clearly these could not have been possible without some knowledge of geometry and surveying. Surveying possibly predated the construction of these buildings since it would have been required for land management purposes of a primarily agricultural economy. Both the Egyptians and the Mesopotamians introduced standard lengths and weights for use in land surveying and in commerce. The units of length were based on the human body—the cubit, the span, the feet. The units of weight were similarly based on the weight of agricultural produce, namely the cereals in use. There is evidence of standardized weights made of stone from both Egypt and Mesopotamia. Balances to weigh were in use in both places—these were equal-armed scales, balanced on a fulcrum.

The Great Pyramid of Cheops

Largest of the pyramids at Giza, this pyramid is also the oldest—built around 2560 BC. The construction of the pyramids in the middle of third millennium BC is still a bit of a mystery since largely flint tools were being used in this period. It is believed that it took 30 years, 100,000 slaves and 2.3 million stones to build the pyramid. (Egypt; Old kingdom/Fourth dynasty)

The rhythm of an agricultural society imposed a form of time measurement on both societies. The Egyptians used the annual flooding of the Nile—an event of considerable importance to the economy—as well as the motion of heavenly bodies to obtain a fairly accurate calendar. The Mesopotamians made several advances in mathematics and astronomy. The sexagesimal system for time and angles as well as the zodiac signs owe their origin to the Mesopotamians. There are numerous examples of the recording of the motions of the stars, planets and the moon on clay tablets. The solar year and the lunar month as well as the division of the week into seven days were all Mesopotamian innovations.

Pottery

Pottery, which had hitherto been made by either molding or by adding successive rings of clay, was to see another innovation sometime in the middle of 4000 BC. A rudimentary potter's wheel started being used around this time—it was simply a turntable which rotated around a pivot, the rotation being provided by turning the turntable with a stick or by hand. This simple invention speeded up the production of pottery tremendously and also paved the way for further improvements in the design to result in a proper potter's wheel many centuries later.

The fact that the same design was used by people around the world over many centuries indicates that the technology disseminated from one single place, probably Mesopotamia.

In this period, there are representations of boats being used for river transport from both Mesopotamia and Egypt. The boats were essentially made of reeds or papyrus and were used for transporting men and materials. There is some indication that the vessels had sails (possibly made from linen), and by 3000 BC, they were essentially paddle boats. The nature of the rivers in the two regions made these light vessels useful for transport and trade. It is only around 2500 BC that there is evidence of rowboats in Egypt. Now wood started being used for construction, leading to much sturdier and larger boats. From 2000 BC onward, the rudimentary sailing vessels, which had evolved from reed boats or rafts, gave way to wooden ships which were oared and also had sails, a high bow and stern. The vessels during this period became much more seaworthy and thus facilitated trade and colonization in the Mediterranean regions.

The period from 2000 BC was characterized by several very important innovations which were to have a profound impact on human society. By this time, the Mesopotamian and the Egyptian empires had consolidated and were well-established centralized states. The demands of the increasing population as well as the need for raw materials like copper, tin and wood would have meant an increasing interaction, either through trade or military means, with the inhabitants of other regions. The areas north of the Mesopotamian region also had contacts with the nomadic people of the steppes of central Asia. This interaction led to an interesting synergy. For instance, the herdsmen of the steppes had domesticated the horse. To this, the Mesopotamian expertise in wood-making was applied to devise the chariot—a vehicle which was to have a major impact historically because of the crucial role played by it in warfare.

Agriculture and Civilization

The settled communities in the river valley of the Nile and the Mesopotamian region grew in number over this time, leading to profound changes in the political organization and control. The Sumerians in Mesopotamia had a well-organized city life—there were temples and houses, and the mainstay of the economy was agriculture with stock breeding, fishing and date palm cultivation. These supported the whole population which included specialists like potters, carpenters, metal smiths as well as the bureaucracy.

The intensive agriculture in Mesopotamia was made possible by canals from the river channels

A relief showing the use of chariot

The introduction of the chariot depended upon several innovations. The bridle and bit had to be introduced for controlling the horses. The wheel was no longer made of solid wood but used a hub and spoke arrangement. Around 1200 BC, the position of the axle was shifted to the rear and the four-spoke wheel gave way to the six-spoke wheel. The yoke which had been used with the ox or with the onager was adapted for use for the horses. Chariots revolutionized warfare and were to have a profound impact on history. In this detail of a relief from the tomb of Tutankhamun, the king is shown hunting animals from his chariot. (Egypt; New kingdom/1370–52 BC; Wood covered with sheet of gold; Egyptian National Museum, Cairo)

Transportation

Transport was revolutionized with the introduction of the wheel. The exact origin of the wheel is not known but from around 3500 BC, there are representations of wheeled transport on pottery and other forms of art. The wheel was first made from wood—specifically, three pieces of wood joined together.

roviding water to land far from the rivers. Initially, when there was less pressure on the land, rivers could have been dammed to irrigate the nearby fields. At a later date, when the cultivation of fruits and vegetables became significant, this would have entailed irrigation throughout the year. In both Mesopotamia and Egypt, a simple water lifting device called shaduf was introduced. This was a long beam pivoted on a pole, with a rope on one end and a water lifting bag or vessel on the other. This simple device increased the efficiency of irrigation and in various forms continued to be used for several centuries.

In addition, draught animals like the ox and the onager were used for plowing. The animals were tied by a rope to their horns, though a yoke was introduced at some point in this period. The use of yokes permitted efficient pairing of animals for plowing as well as for wheeled transport. The earliest plow was a forked stick which was attached to a rope pulled either by humans or by animals. Plowing increased the productivity of the soil. By around 3000 BC, a major modification was made to the plow—the share and the sole were amalgamated into one piece which was attached to the yoked animals. This led to vast improvements in the crop yields. The nutrients in the soil could be recycled and weeding was more efficient since crops were arranged along rows. In Mesopotamia, between 3000 and 2000 BC, a seed drill was added into the sole, leading to further improvement in the efficiency of the plow.

Barley and date palms were cultivated extensively, and both of them were used to make fermented drinks. Beer-making, which involved allowing the barley to germinate and then ferment, is depicted in the tomb paintings of Egypt and Mesopotamian seals dated around 2500 BC. Since the fruit of the date palm is very rich in sugar, the hot climate of these regions made it easy to ferment. Grapes were also cultivated in Sumer and Egypt in the third millennium BC, though the use

of wine was possibly restricted to the royals and the wealthy.

Oilseeds were pressed for oil using very simple oil presses. The simplest one was just a bag which was twisted between two vertical posts to extract oil from olives and even from herbs and other plant products. There is evidence of the use of unguents as cosmetics which were made from oils with sweet-smelling flavors extracted from flowers.

Around 3500 BC, the Egyptians and the Mesopotamians both used a simple form of a loom to weave cloth. The loom was set up on the ground and produced a coarse cloth. Cotton was possibly exported from the Indus Valley to Mesopotamia, though the Egyptians soon started growing it. The choice of fiber in Egypt though was linen, which was made from the flax. In other areas like Anatolia and Syria, wool was also used. Interestingly, many thousands of miles away, in China, in the period around 3000 BC, the evidence of woven silk cloth has been found.

Bread and beer-making in Egypt

Bread-making was common in Egypt and it is conjectured that the same techniques were also used in beer-making. The cereal (barley) was allowed to germinate after moistening. The malted cereal was made into loaves and baked. The baked loaves were then crumbled and allowed to ferment with some water for a few days. The resulting fermented beer was strained and bottled. This painting from a tomb shows bakers mixing and kneading dough and filling bread molds. (Egypt; 18th dynasty/ 1550–1295 BC; Painting in the tomb of Kenamun)

Depiction of the early plow

Agriculture formed the mainstay of the Egyptian civilization with the Nile being the lifeline for both farming and transportation. The early form of the plow was just a forked stick with the junction sharpened to serve as a share. A rope was attached just above the junction and the animals pulled the plow. Later, the rope was replaced by a draught pole secured to a yoke and this was the design which survived for several centuries. In this detail of a vignette from the 21st dynasty, there are details of plowing, sowing and reaping of wheat. (Egypt; *Book of the Dead of Lady Cheritwebeshet*; 21st dynasty/1069–945 BC; Egyptian Museum, Cairo)

Heralding the Iron Age

The techniques of working with bronze also underwent tremendous improvements. Sometime around 2000 BC, bellows were introduced which made possible a much larger scale of metal production than before. Metal, which hitherto had been used only for weapons and some implements, now became common, and there are many more bronze artifacts from this period, exemplifying a much greater variety in casting. Casting technology and metallurgy also improved, and by 1500 BC, bronze-making was very much refined.

While bronze-making was prevalent in Mesopotamia, Egypt and other places, the Hittites in eastern Turkey discovered the use of another metal which proved to be as significant as copper and tin. The Hittites had access to a supply of iron ore. Iron, unlike the other metals which were known then, could not be melted easily given the technology of the furnaces available. However, it was being used in some form or the other.

Working with iron entailed totally different technologies than working with copper or bronze.

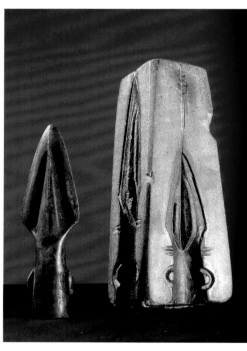

Bronze spearhead and mold

Bronze was discovered around 3000 BC presumably by accident of mixing tin and copper ores. Although bronze was used extensively in Mesopotamia, the situation in Egypt was different. There were large copper mines in Egypt, but due to virtual absence of tin it took another millennium for bronze to become common in Egypt. Improvements in casting and metallurgy refined bronze-making around 1500 BC as can be seen in this image of a spearhead and its mold. (Britain; Early Celtic; British Museum)

The use of bellows did provide furnaces with a higher temperature but the reduction of the metal from the ore was by the gases produced from burning of charcoal. Thus the conditions in the furnace had to be more controlled than with copper. Further, iron being much less malleable than copper, it had to be forged while still red hot. This meant a host of technologies had to be developed before any significant iron-making could take place. These included tongs to hold the hot metal, anvils and heavy hammers.

Iron was being used to make weapons throughout the second millennium BC, especially from around 1500 BC. However, it was only around the turn of the millennium that the established empires of Egypt and Mesopotamia started using iron in any significant quantities for tools. Thus throughout the second millennium, bronze was being used extensively. Craftsmen had discovered ways of casting very large objects by using lost-wax casting in which a model is made from wax and covered with clay. The clay was then fired to remove the wax and molten metal poured into the hollow.

ANCIENT GREECE

This was a period also of the rise of civilization in the Mediterranean, specifically the Aegean Sea. The developments in the Aegean were significant since these gave rise to the Greek civilization in the first millennium BC, which eventually became the first European civilization to leave a distinct mark on the world history.

Like Egypt and Mesopotamia, the Aegean region comprising the islands of Crete, the Cyclades and the Greek mainland, was also developing technologically. In this region, agriculture was being practised since 7000 BC and there were agricultural communities throughout the region. Wheat, barley, oats, lentils, peas and grapes were being grown in the Neolithic Age.

Use of Metal

In the Bronze Age, around 3000 BC, there is evidence of extensive use of metal tools. These included axes—also double axes found in Crete—and daggers. Interestingly, the people used elaborate tweezers for plucking face hair as well as stone palettes for grinding face paint! Also, there is evidence of very elaborate jewelry made of gold and silver.

In 2000 BC, there was an influx of people into this region. Whether these were pastoralists from the north or elsewhere is still debated. However, the mainland experienced some kind of an upheaval around 2000 BC, resulting in the slowdown of technological sophistication with much less use of metals. By 1600 BC though, the region became prosperous again. The materials buried in the graves of this period are very elaborate—gold and silver cups, jewelry and ornaments; there are beads of amethyst from Egypt and ornaments made from amber from northern Europe. Though the weapons are mostly of bronze, there are gold-plated hilts on swords in some cases.

By 500 BC, iron became the metal of choice for tools. As the technology for making iron became more prevalent, iron was used for a variety of things apart from weapons. In military technology, the Assyrians devised the battering ram which proved

Acropolis: the Parthenon

Acropolis, which literally means the edge of a town, was usually a citadel situated on high ground for defensive purposes. The most famous one at Athens had several buildings constructed on it. The Parthenon, the most famous of the surviving buildings, was a temple to Athena built in the fifth century BC. The magnificent Parthenon is considered to be the most important surviving building of the classical period of the Greek civilization.

to be a very effective weapon against the defensive mud and stone walls of cities.

Architecture

The houses were large, with the biggest ones being two-story. On the mainland, the earliest example of the use of baked clay tiles for roof tiling has been found. As early as 1500 BC, there is evidence of well planned cities in Crete with wide squares. The roads were cobbled and the waste water was carried in covered drains. Clay water pipes were also in use, for instance in the palace of Knossos.

Pottery

Building on the technical expertise of the Corinthian pottery workers, whose pottery was valued all over, there were several innovations in the manufacture of pottery. In Crete, a form of the fast potter's wheel was in use by the early Bronze Age, giving rise to exquisite pottery. The potter's wheel was replaced by a much heavier and steadier wheel with the fly wheel placed about 2 ft (60 cm) above the ground. In 600 BC, the wheel head was raised so that the potter could kick the wheel himself while being seated. The first lathe was also used around this time, presumably the inspiration being the Corinthian potters who used to shave the pottery while turning it.

Trade and Transport

The use of seals—made from clay, bone, stone and elephant tusk—and other metal working technologies, including jewelry, was evidently borrowed from Egypt and Mesopotamia. Clearly then, there was contact and trade between the regions. Indeed, there is evidence in the form of paintings on vases of ships which seem to be seaworthy—they are single-masted and have a high prow and a low stern.

As the flourishing civilization traded heavily with its neighbors, the sea transport underwent many innovations. Thales of Miletus, who lived in 600 BC and has been called the father of science, was the first to use triangulation to measure distances. It is possible that triangulation was used earlier by the seafaring Phoenicians from the Levant or the Egyptians whose practical knowledge of geometrical principles was very advanced. Thales is also credited with suggesting the use of Ursa Minor to navigators and was supposed to have predicted the solar eclipse in 585 BC. His pupil Anaximander produced the first map of the world as it was known then. Once again, it is entirely possible that maps were used earlier by the Egyptians and the Babylonians but they were usually local. It was during this time that the triremes—ships with three banks of oars—were introduced in the eastern Mediterranean and then copied by the Greeks. These and other innovations in shipping allowed the Greek soldiers to defeat the much larger Persian force at sea and retain hegemony in the eastern part of the Mediterranean.

The Persians, who had defeated the Babylonians in the sixth century BC, had established an empire from Iran to the Mediterranean. This large geographical entity demanded that communication be faster and hence roads were built all across the empire. The main roads were possibly paved to allow for faster movement of cavalry and infantry.

A ceramic vase

The Greeks made huge advances in ceramics. The Corinthian and Attic wares produced were by far the best in the Western world. The distinctive black and red colors were produced by a very sophisticated process using extremely fine clay and an elaborate sequence of firing. This black-figure hydria from sixth century BC depicts Nereids mourning Achilles. (Greece; 560–550 BC; Louvre, Paris, France)

Mathematics

Alexander of Macedonia emerged as the leader of the Greeks at the young age of 20 in 336 BC. With a desire to rule the world, Alexander conquered much of the ancient world—including the Persian Empire—up to the Indus Valley. After his death in 323 BC, Alexander's huge empire was divided up into smaller parts controlled by his generals. Ptolemy, who controlled Egypt, established a center of learning at the Mediterranean port city of Alexandria, which became famous throughout the ancient world for its museum and library. Among the scholars associated with the library was a pupil of Plato, Euclid. Euclid wrote what has been called the most famous textbook in mathematics: the *Elements*, where he laid the foundations of geometry. Many of the theorems proved in the *Elements* were known

before Euclid but it was only with him tha logical reasoning became established as th method for proving mathematical results. H also studied conics and was responsible for mar results in number theory. Aristotle before hir had laid down the foundations of logic whic formed the backbone of mathematics an philosophy for centuries. Pythagoras Eratosthenes and Archimedes wer some of the other Greeks wh contributed to geometry an number theory.

Other Inventions

Olive oil and wine wer some of the majo exports from Greece The rudimentary press fo pressing olives and grapes wa replaced by a beam press aroun 600 BC. The growing importance c trade in the economy led to the replacemen of the barter system by currency. Coinage wa possibly introduced by Croesus of Lydia. Coinag

ARCHIMEDES: THE GREEK GENIUS

Widely recognized as the leading scientist of the Classical Greek period (490–323 BC), Archimedes was a mathematical and engineering genius. Born in 287 BC in Syracuse, Sicily, he spent some years as a young man in Alexandria where he is known to have befriended Eratosthenes. He was later at the court of King Hieron II where he discovered the principle

of buoyancy. The anecdote about his being given crown to determine if it was pure gold or not an him solving the problem in a bath tub and runnin naked in the streets shouting "Eureka, Eureka" i well known.

Archimedes was an inventive engineer. H improved the screw, which was already being use in Egypt and Mesopotamia, to develop what ha come to be known as the Archimedes' Screw. Thi contraption was very useful in lifting water from canals or even evacuating water from ships.

The great scientist also designed ships for th king, though it is not clear whether they were eve built. There are many stories about his using mirror to burn the Roman ships which were laying siege o Syracuse, though whether these parabolic mirror were actually constructed is not proven. Anothe military device attributed to him is the Claw o Archimedes—a huge crane with a hook to pic up ships and drop them. Once again, there is n evidence of this device having been built.

Archimedes has been called one of the foremos mathematicians of all times. His contributions t mathematics were in diverse fields like geometry infinite series, conic sections and number theory He found the value of the constant pi to a very goo approximation as also the value of the square roo of 3, both irrational numbers which do not have an exact values. He calculated the volumes and are of various solids like cones, cylinders, cones insid cylinders and so on.

Depiction of the Archimedes' Screw

implied that very accurate balances were available to determine the exact weight of the gold and silver being used. It also meant that metallurgical techniques to extract pure gold and silver were developed; in particular, the process of heating impure silver with other substances in a furnace or a crucible was perfected to obtain pure silver.

Greek Influence

As a result of Alexander's conquests, Greek-speaking kingdoms were established in Egypt, Syria, Persia and Bactria, the most prominent being Alexandria in Egypt. A new culture, known as Hellenistic civilization, which was a blend of Greek, Persian and Egyptian influences, emerged after the death of Alexander. By 800 BC, the Hellenes settled in the Aegean region and there were several innovations in science, medicine, technology as well as art in the following eight or nine centuries. Ancient Greece was tremendously influential in terms of shaping Western civilization, its impact continuing through the Roman Empire and down to the Renaissance. Art, culture, literature, politics, philosophy and science are some of the areas where Greek influence can be seen. For instance, the tradition of Olympic Games was started in 776 BC when the first Games were held in Olympia. Western philosophical thought was for centuries almost totally derived from the works of the Greek philosophers like Aristotle and Plato. Aristotelian theory of the heavens or the Greek theory of humors in medicine held sway for almost two millennia before being challenged.

ANCIENT ROME

The period after the death of Alexander, from roughly 300 BC, saw the beginning of the ascendancy of the Roman Empire which essentially lasted till the fifth century AD. Owing to its central location in the Mediterranean, the Roman Empire could easily extend from the Middle East to the British Isles. The empire was governed by a form of democracy for some time, and public infrastructure like paved roads, public buildings, sports complexes and aqueducts were constructed. This period did not see any revolutionary developments, though existing techniques and devices were improved to be used on a large scale.

Improvisations and Innovations

The tradition of innovation at Alexandria was continued with people like Hero and his lesser known contemporary Ctesibius. Hero developed the steam turbine and the water clock which was more accurate than the clocks in use. Ctesibius developed a water organ and a fire engine which was a double

action pump. There were also many improvements in the design of astronomical instruments.

In military technology, Romans improved the weapons like the catapult and the crossbow. They also invented one weapon which used air under pressure. Another new development during this period was in the field of glass blowing, which progressed because of the availability of iron blow tubes.

Engineering and Construction

The improvements in the techniques and devices helped in the construction of buildings and monuments during the Roman times. With the help of devices like cogwheels, pulleys, levers and gears, the Romans could produce better cranes and other equipment which allowed the lifting and movement of heavy weights.

The Roman Empire was a vast one, and the period till the first century AD saw massive construction of roads and aqueducts to serve the empire. However, the Romans were great improvisers, improving upon existing technologies and design but rarely making any new inventions. The aqueducts for instance had been in use in Egypt, Mesopotamia and Greece but the scale of Roman aqueducts

The School of Athens (detail)

Socrates, Plato and Aristotle were the three key figures of ancient Greek philosophy who between them laid the foundations of western culture and philosophy. Plato founded the Academy at Athens which became the first and leading institution of learning in the West in its time. The Greek philosophers discussed a range of subjects like philosophy, music, art, politics, mathematics and science. In this 16th century AD fresco from the Vatican, Plato and Aristotle are seen discussing a philosophical point. (Raphael; 1510–11; Vatican Museums and Galleries, Vatican City)

was much larger. The quality of public sanitation in the Roman Empire was very good—lead and earthenware pipes were used for plumbing. The use of cement in building technology, which led to the invention of concrete, was introduced by the Romans. They designed the brick and concrete arch, replacing the pillar and lintel architecture favored by the Greeks.

The other technology which developed during this time was the water mill. Though the water mills were used around 100 BC in Greece, the Romans expanded their use. Initially, horizontal wheels were used with milling stones to grind corn and wheat. These could only be used where the flow of water was adequate. Then a vertical wheel, based on the water-raising wheel used in Egypt earlier, came into use.

Alphabet

Phoenicians adopted the Sumerian cuneiform and Egyptian hieroglyphics—both using symbols for words—to create 22 letters for spoken sounds. The Phoenician alphabet was modified by the Greeks first and then the Romans to become the modern-day alphabet. Roman numerals are also an adaptation of Etruscan numerals.

Other Inventions

There were many other inventions during this time like under-floor heating used in Roman buildings or the vacuum flask which was found among the ruins of Pompeii. Pliny, the Roman chronicler and scientist, also reports of a method to manufacture soap from tallow and ashes, though this was an improvement upon the basic Babylonian idea of using vegetable oils and alkali salts. The beam press was replaced by the screw press for extracting olive oil. Although there was erroneous theoretical understanding of medicine, the quality of some of the Roman surgical instruments was excellent.

The Roman Legacy

The Roman Empire, though vast, was remarkable in the uniformity in certain respects. Latin was the common language while the Roman legions were recruited from all over the empire. The common law and the Greco-Roman cultural heritage were the other binding factors among the diverse populations. The empire was threatened by barbarians around 150 AD and there was a period of anarchy before Constantine in 324 AD reunited the divergent parts. He also moved the center of the empire from Rome to Constantinople, giving rise to the Byzantine Empire later. Byzantine scholars carried forward the legacy of the Greeks and among their major achievements was the use of mathematics to construct the magnificent Hagia Sofia in the sixth century.

THE ISLAMIC REVOLUTION

The seventh century saw the birth of Islam in the Levant and the establishment of the Caliphate. As the Islamic Empire spread through Spain, North Africa and Persia into the Indian subcontinent, Baghdad—the capital of the Abbasid Caliphate—became the paramount center for learning in the world. The Arabs translated many original works of other civilizations into Arabic and spread this

Roman aqueduct

The brick and concrete arch was among the biggest contributions of the Romans to architecture. Cement was used for the first time by the Romans to make concrete which was used with bricks to make a solid arch. The most magnificent examples of these are in the Roman aqueducts. This aqueduct at Le Pont du Gard, France, built by Romans in 19 BC, was used to carry water across the River Gard.

nowledge through books—they had learnt to make
aper from China. This synthesis of knowledge from
arious civilizations led to the Islamic Golden Age,
vhich lasted from about 750 to 1258 AD when the
Mongols invaded the Abbasid Empire. During this
period, Muslim scholars made several significant
ontributions in the fields of science, mathematics,
medicine and astronomy.

Medicine

Muslims pioneered the concept of modern
hospitals by establishing the first Islamic hospital
n Damascus in 707 AD. They also developed
pharamacology by establishing drugstores and
ompiling encyclopedias of medicine. In the ninth
century, the medical school at the University of
Jundishapur in Persia—incorporating medical
practices of many ancient cultures—became the
argest in the Islamic Empire.

Al-Razi, a ninth century Persian physician,
wrote the *Comprehensive Book on Medicine*
which recorded his experiences as a healer and
provided a useful record of diseases and cures. He
developed treatments for smallpox and measles
and used opium as an anesthetic for the first time.
Many surgical instruments were also invented
around this time, such as the scalpel, the surgical
needle and curette.

In the 10th century, Al-Zahrawi wrote the
extremely influential 30-volume book which
became a compendium of medical knowledge for
centuries. Considered the father of surgery, he
combined the Islamic medicine with the traditional
Greek and Roman medical knowledge. Abu Ali ibn
Sina, who lived in the 10th and 11th century, was
another influential scholar who wrote more than
450 books on a variety of subjects like alchemy
and philosophy. The first person to discuss the

contagious nature of diseases like tuberculosis,
he is best known for his book on medicine which
combined Islamic, Greek and Indian systems of
medicine. The application of mercurial compounds
and purified alcohol as antiseptics was also
practised at this time.

Science, Mathematics and Astronomy

The scientific method was used by scholars,
especially Ibn al-Haytham in the 10th and 11th
century, a remarkable figure who first advocated
the experimental and quantitative methods. His
treatise on optics contained many new ideas
including that of the pinhole camera.

The Muslim scholars developed the mathematical
concepts of the Greeks, Egyptians, Indians and
Babylonians. Muhammad ibn Musa al-Khwarizmi
wrote a book on algebra which gave the first
systematic solution of linear and quadratic
equations. These and other equations had been
discussed much earlier by the Greek mathematician
Diophantus but Khwarizmi was the first person to
develop it as a discipline.

Abu Musa Jabir ibn Hayyan—called the father
of chemistry—made many important discoveries in
the eighth century. He introduced the experimental
method in alchemy and invented many chemical
processes—the synthesis of nitric and hydrochloric
acids being the most prominent one. Distillation
process was made more sophisticated at this
time—alcohol and petroleum were some of the
products which were distilled.

Al-Farghani, Al Sufi, Al-Zarqali and Al-Bitruji
were the famous Islamic astronomers of this period
who challenged the existing astronomical theories.
Among several achievements in astronomy, the
Muslims built the world's first observatory and
invented the astrolabe.

Influence on Europe

The Golden Age had a tremendous influence on
the development of science and technology in
Europe in the subsequent centuries. The knowledge
accumulated during these five centuries proved to
be extremely useful in a variety of fields of both
applied and pure science. The translations into
Arabic of many of the classical Greek texts which
had been lost were invaluable to the medieval
scholars in Europe.

The Islamic scholars were also instrumental
in bringing the scientific and technological
developments of India and China to the notice of
the European scholars. The progress in the Middle
East continued for another few centuries after the
end of the first millennium, resulting in many
new technologies and the rise of new institutions
like the University as the new site for research
and knowledge.

OTHER ANCIENT CULTURES

ANCIENT CHINA

The period from 4000 BC to 1000 AD was also the time when another great civilization was born and prospered far away from the Levant and Europe. The river valleys of Huang Ho and Yangtze rivers saw the birth and maturing of one of the most technologically advanced civilization—China. As early as second millennium BC, the Chinese had a highly developed writing system.

The first millennium BC saw the emergence of the Chinese practice of acupuncture for the treatment of various diseases, though some form of acupuncture was practised even in the early Bronze Age. Chinese medicine, which developed independently of the Greek and Indian traditions, was codified toward the end of the first millennium BC.

The Chinese, like other ancient cultures, were very interested in astronomy and astrology. They developed improved sundials to keep time, and sometime in the early first millennium BC, developed the abacus for counting. The first systematic observations of comets as well as detailed descriptions of planetary positions are recorded during this period.

mention of the use of saltpeter (potassium nitrate) charcoal and sulfur to manufacture the substance Gunpowder, along with paper, printing and the magnetic compass, comprise the "four great inventions" of ancient China.

Although papyrus was used by the Egyptian much earlier, the Chinese were the first to make true paper from pulp around 100 AD. This was made by Cai Lun, a court official, with mulberry bark old rags and other fibers. In a couple of hundred years, its use became widespread for writing. By the seventh to eighth century AD, paper bags were used for storing tea and as toilet paper. Paper-making reached central Asia by the eighth century and soon was prevalent all over the Caliphate.

Printing, the other great invention, seems to have been developed by the Japanese who used stone blocks to print prayer books in 760 AD. But it was the Chinese who by the ninth century had perfected the art of woodblock printing and, by the turn of the millennium, had even experimented with some kind of movable type made from ceramic. The type was unsuitable since the number of characters in the Chinese language were too many! The *Diamond Sutra* was the first printed book. Made from seven strips of paper joined together with an illustration on the

A woodblock printed version of the *Diamond Sutra*

The *Diamond Sutra*—the world's earliest surviving printed book—consists of individual sheets of printed text and a frontispiece illustrating the Buddha surrounded by acolytes and disciples. (China; Tang dynasty/868 AD; British Library, London)

In the field of military technology, the Chinese had developed and perfected several weapons many centuries before their appearance in the West. The crossbow was used by the Qin dynasty in the third century BC, and many were buried with the famous terracotta army in the tomb of the Qin emperor.

The military invention which changed the course of human history—gunpowder—was developed in the fourth century AD. There are indications of experimenting with various chemicals similar to those used for gunpowder in earlier texts, but it was in 300 AD that there was the first definitive

first sheet, the 16-ft (5-m) long scroll was printed in 868 AD. Being the first to introduce paper and printing, the Chinese also invented paper money as early as the ninth century AD.

In the field of metallurgy, specifically iron-making, the Chinese took a totally different route than the other cultures. Elsewhere, iron ore was first reduced to a bloom which was then hammered into wrought iron. The hammering forced the impurities out of the bloom. In contrast, the Chinese went straight from the iron ore to cast iron, that is, molten metal which could be cast into molds. This was possible since the iron ore available in China had a relatively lower melting point and also because the Chinese had developed a form of piston bellows which could produce a steady draught in the furnace made from good quality clay.

The Chinese knew a lot about magnetism by 500 BC. They knew that magnetitite, an iron ore, aligned in a north/south position; they also knew how to make magnets. By 1000 AD, compasses were widely used for navigation on Chinese ships. Other early Chinese innovations include movable sails and rudders in ships, porcelain, canals, roads and inns.

INDUS VALLEY

Sometime during the middle of the third millennium BC, there emerged yet another great civilization in the region east of the Iranian plateau along the Indus River. Though there were settled communities in this region around 5000 BC or even before—Mehrgarh in present day Baluchistan was a thriving settlement which has been excavated—the Indus Valley Civilization emerged as a mature culture between 2600 and 2500 BC.

The ancient cities of Harappa and Mohenjodaro, excavated in the early 20th century, indicate an amazingly developed urban culture—well planned cities with a huge citadel, streets laid on a rectangular grid and houses with advanced sanitation systems. Most houses had private wells and bathrooms from where waste water went into covered drains which lined the streets. Stone was rare because of non-availability of quarries in the vicinity. Bricks—either fired or sun dried—were standardized and joined in a fashion which has been used since then for centuries.

Agriculture continued to be the mainstay of the people with new varieties of wheat and barley being grown. In addition, date palm, sesame seeds, melons and peas were also grown. Mustard was cultivated, as was cotton which was used for textiles and also exported. The Indian humped cattle, sheep, goat, camel, fowl, cat and ass were some of the animals who were domesticated.

The Indus Valley culture was spread over a fairly large area and its homogeneity suggests that communication between the various regions was efficient. In addition, there was booming trade with Mesopotamia as is evident from the presence of seals from the Indus valley cultures found in Mesopotamia. Trade with the Middle East was carried out by the overland route or by seaworthy vessels, though in the absence of any evidence of navigation aids, they possibly stayed close to the coast.

The Indus Valley people developed a uniform system of weights and measures. They also had a written language, which remains undeciphered to the present day. There is evidence of proto-dentistry, stringed musical instruments, and various toys and games from the valley. In addition, there is plenty of evidence of crafts—bronze and copper were being worked into implements like chisels, axes and also in objects of art like figurines. The other metals in use were lead, gold and silver. Textiles were made from cotton and there is presence of dyeing vats in

the excavations, indicating a fairly advanced textile industry. Bead-making and faience industry was also quite developed. Many of the raw materials used by the craftsmen were not available locally and hence there must have been extensive trade with other regions of the world.

The Indus valley civilization lasted for around 1000 years before the nomadic raiders from central Asia, called Aryans, invaded India and established a new culture.

ANCIENT INDIA

In the period from 1500 BC to 1000 AD, essentially the whole of the Indian subcontinent was settled and there were great advances in the fields of astronomy, grammar, mathematics, medicine and other areas. India's classical period comprises the reigns of the Mauryans and the Guptas in the third and fourth centuries, a period of great cultural and technological achievements.

There are many treatises on astronomy and astrology from India. The earliest reference to the heliocentric model of the universe is found in a ninth century BC text. In the third century, Yajnavalkya, a sage, proposed a model to explain the motion of the sun and the moon and calculated the length of the year fairly accurately. This tradition of astronomy was carried forward because of the linkages to the Vedic rituals.

The water conduit at Mohenjodaro

The cities of Mohenjodaro and Harappa in the Indus Valley are known for their planned structure and advanced drainage system. Baked bricks were used for all constructions for greater durability. (Harappan civilization; 2500–2000 BC)

In the fifth century AD, the famous astronomer and mathematician Aryabhata gave a mathematical theory of the solar system and correctly stated that the light from the moon is the reflected light of the sun. He also calculated the ratio of the length of the earth's rotations to the lunar orbit.

Mathematics was an area where the Indian scholars contributed many new concepts. The decimal system and the concept of zero are among the most important contributions of Indian mathematicians. In number theory, there are references to square roots, cube roots and algebraic equations. Geometry was a very important discipline in this period since the Vedic rituals involved the use of fire altars for sacrifices. The construction of various kinds of fire altars—with varying dimensions and shapes—and the use of an exact number of bricks for this purpose was extremely critical for the sacrifice to be successful. *Sulba Sutras*, a book written between the seventh and fifth century BC, contains very elaborate methods of construction of altars and the first reference to the Pythagoras Theorem is found in it. Apart from these, there are numerous references to combinatorics, including Pascal's Triangle and Fibonacci numbers, first found in the work of Pingala in the fourth or third century BC.

The Indian system of medicine too was quite advanced by this time. Ayurveda, the science of medicine, was practised from the ninth century BC but it was codified into a treatise by Charaka in the fourth century BC. The pharmacopeia of Ayurveda was based on herbal medicines as well as minerals, and there was a self-consistent theory of medicine. Surgery was also practised in ancient India—with their being some evidence of drilling of teeth from Mehrgarh from 7000 BC—but the diverse traditions and practices were consolidated by Susruta in the sixth century BC. His book on surgery describes many instruments and procedures including those for plastic surgery, especially of the nose, and the removal of the cataract from eyes.

Indian artisans, like their counterparts in other parts of the world, developed many techniques for working with metals like copper, tin and iron. Even as early as the fifth century

BC, iron was being used for weapons. It is believed that the crucible method for making steel was discovered in south India in the fourth century BC. The most famous steel was the wootz steel which was made by heating iron, charcoal and glass in a sealed crucible or furnace. This steel was prized throughout the ancient world and exported to many parts. It was used in Damascus to make the famed Damascus swords which were famous throughout the world for their sharpness and durability. There were many advances in the extraction and refining of metals like gold, silver and mercury in India during this period.

ANCIENT AMERICAN CIVILIZATIONS

Far away, in the thick, tropical jungles of central America, another civilization developed in this period. Domestication of several plants took place in this region around 7000 BC, and by 1500 BC, there were settled agrarian communities growing corn, beans, squash and cotton in the region. The diverse communities were unified into a more uniform and centralized society around 1200 to 900 BC. By 500 BC, this centralized order gave way to regional cultures and kingdoms like the Maya, Zapotec and Totonac.

The Maya civilization lasted till the ninth century AD but was at its peak in the fifth century AD. The Mayans developed a writing system and had a very highly developed calendar and astronomy. They had calculated the duration of the solar year to an amazing accuracy as also the orbital parameters for Venus. For some obscure reason, Venus was more important for the Mayans than even the sun. They had very precise lunar tables and could predict solar eclipses. Their mathematics was also developed with positional values and even a zero. Though the concept of zero was already known in India much before this time, it is evident that the Mayans discovered it independently at a later date.

The Mayans also made a form of paper in the fifth century AD from the inner bark of some kind of a fig tree, which was used to prepare manuscripts, a few of which survived the Spanish conquest. The manuscripts contained astronomical tables, dynastic histories and court records.

In the Americas, the most advanced technological developments took place in the southern part of the land mass. In the narrow strip of land between the Andes and the Pacific, in what is today's Peru and Ecuador, there are many rivers flowing down from the Andes. By 1000 BC, there were settled agrarian communities established in these river valleys. Among the many achievements of these people was a highly developed weaving industry with wool from llamas or alpacas.

The other advantage of being near the Andes was the metals available in the mountains. There were deposits of gold, silver, tin and copper in the region.

Jade plaque with Mayan inscription

The earliest known writing from the Maya is from 250 BC. But the script was probably developed much earlier. The writing, comprising hundreds of figures or glyphs, was very complex. The script was either carved in stone or written on wood, ceramics, manuscripts, or jade as shown here. (Mexico; Fifth century AD; Private Collection)

In the seventh or sixth century BC, there was already a gold working tradition in the region. Later, toward the end of the millennium, the Peruvian craftsmen were already melting and casting gold, alloying it with silver and copper while their counterparts in Ecuador were alloying gold and platinum! Still later, in the early centuries of the second millennium AD, the Andean region saw the emergence of another advanced empire: Inca.

ANCIENT JAPAN

The earliest culture known from Japan was the Jomon culture which is dated from 7500 BC to about 250 BC. The early Jomon period saw the manufacture of very sophisticated pottery which included urn-like vessels with complex designs. The habitation in this period was of two kinds—a pit type structure with a roof and a circular floor of clay or stones covered with a roof. There is no evidence of weaving during the early period with clothes mostly made of bark, though there is evidence of ornaments made of seashells, clay, stone, bone and horn. Plants like taro and yams were cultivated at this time, though agriculture was introduced later.

The Yayoi culture arose during the later Jomon period—around third century BC—and this represented major advances in pottery, agriculture and technology. The Yayoi pottery, as compared to the Jomon pottery, was fired at much higher temperatures and turned on a wheel rather than molded by hand; essentially designed for practical use, it was much less decorative than Jomon pottery. It was during this period that cultivation of rice was introduced in Japan. The farming techniques employed for paddy cultivation were quite advanced. The Yayoi used iron and later bronze for making tools. Cloth was now woven on looms and settlements were much like the ones in Jomon period, with the addition of some storage structures for grain.

The Yayoi period was followed by the Tumulus period which lasted from about 250 AD to 550 AD and witnessed the introduction of Buddhism in Japan. After the Tumulus period was the Age of Reform, followed by the Nara period and finally the Heian period which lasted till the 12th century AD.

OTHER CULTURES

While the achievements of these cultures are well documented and the archeological sites provide enough evidence, there is another set of people whose achievements in certain fields were as remarkable. Foremost among these are the nomads of the central Asian steppes. These people domesticated the horse and in all likelihood made the first wagons pulled by a pair of draught oxen. The wagon, which led to the development of the chariot, played an important role in the military history of humankind. Another great invention of the nomadic people was the stirrup. There are indications that as early as the fifth century BC, stirrups were being used by the nomads in central Asia. The stirrup allowed the rider to use his weapons more effectively and greatly improved the military potential of the cavalry in the coming years.

In addition to these, other ancient civilizations— such as the Etruscan, the Hebrews, Phoenicia, Persia and Kush—left a mark on the world with their wide-ranging inventions.

Mayan calendar

The Mayans are credited with developing an accurate calendar. There are a series of dates around the edge of the disc shown here, representing the calendar developed by the Maya civilization. (Mexico; 590 AD; Stone; National Museum of Anthropology, Mexico)

Dotaku bell

The bronze culture in Japan was restricted to a small area and lasted from the middle Yayoi period to the Tumulus period. Dotakus are slender, bell-shaped bronze forms, inscribed with geometric or lacework patterns. (Japan; Third century BC; Museum of Eastern Asia Art, Cologne)

POST-CLASSICAL TO RENAISSANCE
SCIENTIFIC QUERY AND DEVICES (1000–1400)

The last few centuries preceding the new millennium had seen tremendous progress in technology and science in the Islamic world as well as in far away China. In fact, the period from eighth century AD to the middle of the 13th century has been called the Golden Age of Islamic Science. The situation in China was also similar around this time. And, contrary to what is usually thought, there were many changes and developments in Europe as well in this period.

GOLDEN AGE OF ISLAMIC SCIENCE

The centers of learning in the Islamic world included Baghdad (as the capital of the Caliphate), Spain and Sicily. At these places, there were many scholars who worked on a variety of fields and made significant contributions. Ibn al-Haytham, a remarkable polymath, worked on astronomy, physics and anatomy among other areas. His *Book of Optics,* written in the early years of the 11th century, was the first comprehensive treatise on optics and contained many fascinating discoveries. He enunciated a theory of vision and also carried out many experiments with lenses and mirrors. This book was remarkable in that it talked about

many laws and phenomena of optics which were rediscovered independently many centuries later such as the laws of refraction and the phenomenon of dispersion of white light into its constituent colors. In astronomy, he conjectured on the theory of gravity as well as the shortcomings of the dominant Ptolemaic geocentric view of the universe, though this could have stemmed from his familiarity with the work of Indian astronomers.

Astronomy

In the eighth century, brass astrolabes were invented in Persia. An astrolabe was a basic instrument used by astronomers to determine the position of the sun, moon and the stars. It was a navigation aid and also used extensively in surveying. The first astrolabe was possibly made by the Greek astronomer Hipparchus in the first century BC and was improved during the intervening period. The importance of the astrolabe was even more in the Islamic world since it was used to find the local time to determine the schedule for the daily prayers. It was also used to identify the direction of Mecca, the direction in which Muslims are supposed to face during their prayers. Brass astrolabes were immensely useful since they were accurate and could be transported easily. It was from Islamic Spain that the astrolabe was

The Aztec Calendar Stone

Human sacrifice formed an integral part of Aztec culture. It is reported for the reconsecration of the Great Pyramid of Tenochtitlan in 1487, more than 80,000 humans were sacrificed. The image is a detail of the great Calendar Stone from the Great Pyramid. At the center is the mask of Tonatiuh crying out for the blood of sacrifice. Surrounding this is a group of symbols depicting the earthquake which will end the world and around this are all the signs of the days of the Aztec year. (Great Pyramid at Tenochtitlan, Mexico; Late postclassic; National Museum of Anthropology, Mexico City)

1000: Erik the Red sails to North America. Mahmud Gahzni begins his raids into India. Emergence of watermills in Europe.

1010: Ibn Sina compiles the *Canon of Medicine*, an extremely influential medical text book in Iran.

1025: The Cholas of South India conquer South East Asian territories.

1054: The Great Schism between the Western Roman Catholic and Eastern Christianity.

1066: Battle of Hastings, leading to Norman conquest of England.

1085: Alfonso VI of Castile captures the city of Toledo.

1095: First Crusade.

1100: Rise of the Inca Empire.

1120: Windmills introduced in Europe.

1160: The University of Paris is separated from the cathedral school of Notre Dame.

1170: University of Oxford founded.

1171: Saladin captures Egypt and poses a challenge to the Crusaders.

1190: Use of magnetic compass in China and in the Islamic world.

1204: Crusaders recapture Constantinople. Arabic numerals introduced in Europe to replace Roman numerals.

1206: Rise to power of Genghis Khan; increased trade and contact between China and Europe.

1215: Magna Carta signed.

1250: Aztecs migrate to central Mexico. First guns appear in China.

1258: Baghdad sacked by the Mongols.

1271: Marco Polo begins his voyage to China.

1271: Beginning of Yuan dynasty in China. Mechanical clocks invented.

1315: The Great Famine in Europe kills millions of people.

1325: Tenochtitlan, the capital of the Aztec Empire, founded.

Water wheels on the Orontes River, Hama, Syria, Middle East

1337: The English and the French wage the Hundred Years' War.

1347: Plague kills vast numbers of people in Europe.

1368: End of the Yuan dynasty in China and beginning of the Ming dynasty.

Science

The field of alchemy and chemistry also saw some major developments in the Islamic world. Abu Ali ibn Sina, a scholar whose work on medicine was the standard book for many centuries, discovered the process of steam distillation in the early part of the 11th century. Al-Biruni made the first laboratory flasks and the specific gravity bottle. The Islamic scientists of this time studied, classified and manufactured many substances and herbs which were used as medicinal drugs.

There was a lot of innovation in technology too around this time. By far the most important figure in this field was Ibn Ismail ibn al-Razzaz al-Jazari. Al-Jazari, who lived in the late 12th and the early 13th century, is considered the father of engineering, thanks to his amazing inventiveness in a variety of engineering fields. The list of devices invented by Al-Jazari is a long one, including automatons, gear mechanisms and water lifting devices. Among his most important invention is the crankshaft—a device which converts the linear motion of a piston into the rotary motion of a wheel. This one device is hailed as the most important innovation in transport since the wheel because it made possible the harnessing of the power of steam and gasoline in the internal combustion engine several centuries later. He also created the connecting rod which, along with other inventions, helped to develop water raising machines.

Al-Jazari was also the first person to talk about a suction pump with a valve—in fact, he invented the two-cylinder reciprocating suction piston pump which was driven by a water wheel. He designed and

introduced to the Europeans in the 11th century. In the 13th century, an astrolabe with a calendar and gears—which is the world's earliest example of gear mechanism—was made by Muhammad . Abi Bakr, and in the next century a spherical astrolabe was also developed.

Abu Rayhan al-Biruni, a renowned polymath in the 10th and 11th century, did some pioneering work in the field of astronomy. He conducted many experiments and described the solar eclipses and lunar eclipses in great detail. He also conjectured that the Milky Way galaxy was not one single object but in fact a conglomeration of numerous stars. He calculated the radius of the earth to an astonishing precision and came up with a figure which is only marginally different from the actual one. He wrote several treatises on a variety of astronomical topics, including the planisphere, astrolabe, astrology and a comparative study of the different calendars used by various cultures.

Mathematics

The mathematicians in this period were aware of the advances made in various fields by the Indians as well as the ancient Greeks. Among them was the famous Omar Khayyam, better known for his poem "Rubaiyat". Khayyam was an accomplished scholar—apart from being a great poet, he was also an astronomer, a mathematician and a philosopher. In the 11th century, he generalized the method of the Indian mathematicians to solve cubic equations by using the intersection of a circle with a conic section. He also discovered the binomial expansion and conjectured about the correctness of the postulates of geometry propounded by Euclid many centuries ago. At the time, there were several advances in other fields of mathematics like number theory, geometry, and even cryptography.

The Caird Astrolabe

Islamic scholars made detailed astronomical charts and used a variety of instruments. The most important instrument was the astrolabe which was used for locating the positions of the heavenly bodies. It was also used by navigators and surveyors. The brass astrolabe was first made in the eighth century by Islamic mathematicians. Astrolabes were used to determine the direction of Mecca and also the timings of sunrise and sunset to schedule the ritual daily prayers. This is an image of the Caird Astrolabe from 14th century Spain. Astrolabes in medieval Europe were used for making horoscopes. (Spain; 14th century; National Maritime Museum, Greenwich)

Vertical scoop wheel

Remarkable progress was made in science and technology during the Islamic period. Apart from translating Greek texts into Arabic, the Islamic scholars were also responsible for several technological innovations. Al-Jazari was an exceptionally talented scholar who made fundamental contributions in many fields. His book *The Book of Knowledge of Ingenious Mechanical Devices* contained more than 50 mechanical inventions together with their detailed drawings. This is a miniature from the chapter "Trick Vessels", which depicts a vertical scoop wheel used for filling wine reservoirs. (Egypt; Mamluk copy, first written 1181–1206)

made many automata which used water wheels and cam shafts together with gears. He is also credited with constructing the first humanoid robot—a set of musicians which could be made to play musical instruments and were powered by water—among many other devices such as a water-powered clock, a combination lock and a segmental gear.

The field of medicine also saw much progress in these times. Building on the tremendous work done in medicine and surgery in the last few centuries, advances were made in understanding and healing the human body. The first description of the human circulatory and pulmonary system was by Ibn al-Nafis in the middle of the 13th century while Mansur ibn Ilyas described the nervous system in the 14th century. There was also investigation into psychology and the effect of the mind on the human body.

Water clock with automated figures

Al-Jazari is credited with many inventions in the Islamic Golden Age. The water clock is mentioned in the "Treaty on Mechanical Procedures" in 1206. Powered by water, this clock went through many modifications in subsequent centuries. (Turkey; 1206; Vellum; Topkapi Palace Museum, Istanbul)

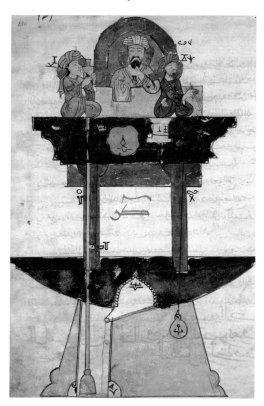

END OF ISLAMIC EMPIRE

The Islamic Empire's downfall began at different points of time due to different causes. The rise of Christianity in Europe impacted the Islamic Empire in Spain, culminating in the recapture of Toledo in 1050. This event proved to be of great importance in the history of Europe since among the first to enter the city were monks who got hold of the huge number of texts and translations of the ancient Greek works in Arabic. These included not only the original works of Islamic scholars but also the translations of the ancient Greek masters. This allowed the European scholars to get access to the ancient Greek scholarship, something

which had been lost during the Roman an Byzantine periods.

In 1206, the world witnessed an event whic was to change the contours of the civilized worl for centuries to come. A Mongol named Gengh Khan established the Mongol Empire, and withi a few years ruled over the largest empire the worl had ever known, extending from the northeaste tip of Asia to well into Europe.

In 1258, Baghdad was sacked by the Mongo and centuries of scholarship were destroyed the process. The Caliphate ended and Mong rule was established, bringing the Golden Age Islamic Science to an end. The next renaissanc of Islamic scholarship came some centuries late at the height of the Ottoman Empire in th erstwhile Byzantium.

CHINESE TECHNOLOGY

The new millennium also saw tremendous change in China. After a brief period of civil wars, th Song Dynasty was established in 960 AD. Thi period is notable in Chinese history for some the most radical technological innovations. Th period was one of stability and the rulers promote meritocracy through a Confucian bureaucrac which was recruited through competitiv examinations. The atmosphere was thus conduciv to new ideas and thinking.

This was also the period of an agricultura revolution with increasing use of improved too and mechanical implements, leading to a highe productivity. Cotton and new varieties of earl ripening rice (imported from Cambodia) wer some of the innovations in agriculture.

Printing
In the 11th century AD, the ceramic movable typ was introduced in printing, to be replaced by th wooden movable type in the 13th century. Th movable type made printing of books possible i much larger numbers. This was a very significan development since now many more people coul prepare for the civil service examinations The Song Dynasty also introduced paper mone and there were several mints established fo printing banknotes.

Navigation and Transportation
Shen Kuo, a Chinese polymath, developed magnetic needle compass in the 11th century. H also was the first person to determine the directio of true north as opposed to the magnetic north making compasses more useful for navigation.

The records of Chinese ships visiting Afric indicate that there were great advances in ship building in the Song dynasty. Features like watertight bulkhead compartments in the hull were introduced; the paddle ships in use for some time were also improved in this period. The ships used for war and trade were huge—fit for carrying large amounts of ammunition and cargo—and manned by hundreds of sailors. There were also large sailing ships—the characteristic Chinese junks—with as many as a dozen sails, made of matted bamboo, which were capable of great speed and mobility.

Military Technology

The Chinese not only invented the gunpowder but were also responsible for several advances in the military technology. As early as the 10th century, flame throwers were used with some kind of flammable material on their enemies from the north. In the 11th century, the Song warriors were throwing "bombs" from catapults.

Bamboo tubes—the first guns that we know of—were used to throw lead pellets, sand and stones. At a later date, the bamboo tubes were replaced by cast iron tubes. Cannons were made with either bronze or cast iron and used to throw heavy lead balls which were filled with gunpowder. The earliest use of handguns is dated from 1288 in China. Land mines were first introduced by the Song dynasty in the late 13th century; these used booby traps which lit the fuses. The Chinese were also the first to make gunpowder-propelled rockets in the 13th century.

Other Inventions

Wind power had been used by the Chinese for a long time, as is obvious from the prevalent bronze casting and cast iron, both of which require huge bellows to obtain the high temperatures needed in the furnaces. Now windmills—possibly inspired by Persia—were introduced in China.

Metallurgy also saw several innovations in the first few centuries of the new millennium. Improved methods of making steel from iron were discovered; in particular, a method using partial decarbonization of iron and repeated forging was devised around this time. In fact, the production of iron and steel was so large that it posed a serious threat to the forests, which were used for fuels. This changed in the 11th century when coke started being used in place of charcoal. Iron was utilized for weapons, cooking utensils, machinery and tools.

In 1088, in the city of Kaifeng, the polymath Su Song made the first hydraulic-powered astronomical clock tower. Human beings had been measuring time for centuries—first by the natural rhythm of the day–night cycle and the longer cycles of seasons, and later by various regular phenomenon like the flow of sand or water. Su

Song's clock was remarkable in that it used an endless chain drive for the first time. Any basic time-measuring device or clock needs to have three parts—a source of power which could be water, falling weights or even a spring as was used in later centuries; an escapement mechanism, something which will allow a slow release of energy; and finally a set of gear wheels to control the speed. Su Song's water-powered clock tower had an escapement mechanism which was used for the first time.

The mechanical ingenuity of the Chinese engineers in this period was remarkable. In the 11th century, they made the first odometer—a device to measure the number of miles traveled by a chariot or a wheeled vehicle. This was a contraption made from a system of interlocking gears with different numbers of teeth. In fact, the first known use of differential gears, of the kind used in modern vehicles, is from this period too. Differential gears are used to regulate wheels which are moving at varying speeds while taking a turn. These were used in the south-pointing chariot which utilized gears to maintain its bearings. Composed of a complex gear mechanism, the figure on top of the south-pointing chariot always pointed in the same direction, thereby acting as a non-magnetic compass.

Exchange with Europe

The Mongols defeated the Song dynasty and established the Yuan Dynasty in 1271. The great Mongol emperor Kublai Khan was the founder of the Yuan dynasty with its capital at the newly constructed city of Beijing. It was during the reign of Kublai Khan that the Venetian explorer Marco Polo came to China and spent several years at his court. Marco Polo played a major role in bringing

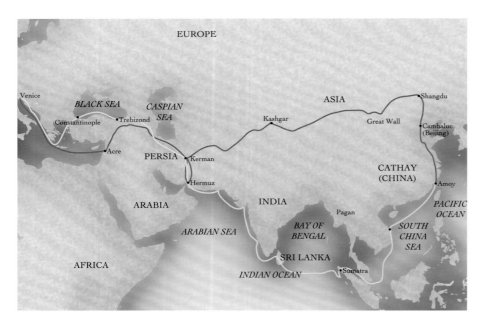

EUROPE
ASIA
Venice
BLACK SEA
CASPIAN
SEA
Constantinople
Trebizond
Kashgar
Great Wall
Acre
PERSIA
Kerman
Hermuz
ARABIA
INDIA
Pagan
ARABIAN SEA
BAY OF
BENGAL
SRI LANKA
INDIAN OCEAN
Sumatra
AFRICA
Shangdu
Cambaluc
(Beijing)
CATHAY
(CHINA)
Amoy
PACIFIC
OCEAN
SOUTH
CHINA
SEA

Marco Polo's route to China

In the 13th century, Marco Polo, a young Venetian, traveled with his father and uncle on the Silk Road—an ancient trade route—to become the first Europeans to visit the capital of China. Kublai Khan, the Mongol ruler of China, appointed them to the royal court, and Marco Polo spent the next 17 years traveling on missions to many parts of Kublai's empire. After returning to Venice via sea, Marco Polo wrote a book about his experiences which became an important source of information during the 14th century about the Middle Kingdom.

The Madrid Codex

The Mayan calendar, like all Mesoamerican calendars, was of 260 days. Other forms of the calendar were used to track longer periods of time. Among them was an unusual 584 day Venus cycle which recorded the conjunctions and appearance of the Venus. Mayan rituals and ceremonies were governed by auspicious times according to the calendar. This image is a detail from the Madrid Codex, one of the four surviving codices. The codex is almost entirely composed of almanacs corresponding to the 260-day ritual calendar used in Mesoamerica for divination and prophecy. (Mexico; 13th century; Paper made of fig tree bark; Museum of Americas, Madrid)

to the notice of the West the fabulous advances in technology in China.

The early Mongol rule was significant since the emperor encouraged trade via the Silk Road, the artery connecting China to the West. This was the channel through which many of the Chinese inventions and technologies reached the West after this period. These included porcelain, printing techniques and other products. In the reverse direction, one of the main imports of China at this time was sorghum from Europe. There were no major innovations during the Yuan dynasty which ended in 1368.

RISE OF MESOAMERICAN CIVILIZATIONS

The first few centuries of the new millennium was also the time when the Americas saw the decline of an old civilization and the birth of two major powers. The Mayan civilization, which had reached its peak around the eighth and ninth century AD, was now in a state of decline. The Mesoamerican region was inhabited by various empires but the level of technology was relatively poor. The basic tools in use were still made of stone since metal was primarily used for weapons and ornaments.

Agriculture was hugely productive despite the use of stone and wooden tools. The primary crops being grown were rice and maize, though this was supplemented by beans to provide the protein and a variety of other vegetables like squashes, peppers, tomatoes, cassava, cotton, tobacco and so on. The reason for the high agricultural productivity was the practice of slash and burn agriculture where the cleared fields were left fallow after a few years. Extensive terracing and irrigation was used in the highland areas.

The remarkable thing about the whole culture was the absence of any draft animals and hence the extensive use of human power. The pyramids, the ceremonial platforms and other structures built during these periods, all used human labor. Stone was the material of choice; woodwork, weaving and metal working with gold and copper were common and widespread.

Aztec Civilization

The decline of the Mayan culture was followed by the rise of Aztec in the 13th century, in what is now Mexico. The Aztecs developed very innovative agricultural techniques for the mountainous habitat of their empire. Soil erosion on the sloping terrain was controlled by terraces made of earth and stone. The Aztecs also pioneered swamp reclamation and used the lakes for agriculture. The fairly large lakes were converted into farmland by an ingenious system of drainage ditches, dikes and sluice gates.

The basic technological level of the Aztecs was not very superior to the Mayans, though the Aztecs managed to build a huge capital city at Tenochtitlan in Mexico in 1325. This city was in the middle of a lake and had important architectural structures like the Main Temple, the Great Pyramid and many famous palaces. The siege of Tenochtitlan in the 16th century by the Spanish conquistadores marked the downfall of the Aztec Empire.

Inca Civilization

In the southern part of the Americas, this period saw the consolidation of a great empire by the Incas in the 13th century. The Andean empire of the Incas was spread over the present day Ecuador, Bolivia, Argentina, Peru and Chile. The diverse habitats, ranging from coastal deserts to the high Andes, were used for agriculture because of the

ome unique features like the many rivers flowing to the Pacific from the Andes and the latitude of the region. Tubers of many kinds were cultivated even at very high altitudes where the land was left fallow for several years after some years of cultivation to allow for rejuvenation of the nutrients. There were also herders using the Andean pastures for raising llamas and alpaca herds.

The extreme variation in the temperatures during the day and night were used very creatively to preserve meat and tubers. In fact, the Quechua name for the processed meat is *charki* which is the origin of the English word "jerky"! The rivers were channeled, the slopes terraced and the valley floors used effectively and efficiently to increase agricultural productivity.

Weaving, of which there was a long tradition in the Andean regions, reached a high point during this period. Painted and woven cloth with exquisite designs was used for burial and sacrifices. There were specialized women weavers to weave the cloth, as textiles were highly valued in the Inca society.

The widespread Inca Empire had a very efficient communication system. The vast network of roads and tracks were important since it allowed

recording data. Records were kept of history, of the number of domestic animals, of textiles and even of imported hens!

The network of roads formed the backbone of the Inca Empire. There were more than

15,000 miles (24,000 km) of highways criss-crossing the whole empire and these were maintained locally. Suspension bridges were another Inca innovation. Made from woven ropes or textiles, these were remarkably sturdy and are to be seen even to this day. In the absence of wheeled transport, these bridges formed an important aspect of infrastructure.

Incan architecture was stone-based and the grandeur of the capital city of Cuzco as well as the ruins of Machu Picchu are testimonies to the ingenuity of Inca stoneworkers. Rediscovered in 1911, the site of Machu Picchu, made of finely-cut gray granite, is an architectural wonder. The stones are made to fit perfectly without the use of mortar or any kind of adhesive.

Bronze tools and weapons were used extensively in the region during this time. The Incans also discovered the use of cocoa leaves to relieve exhaustion, especially among the runners who carried the royal messages across the empires. There is also evidence of them performing some early form of surgery on the skull to relieve cranial pressure.

In the 16th century, the introduction of contagious diseases like smallpox and diphtheria from Europe caused widespread decimation of the local Incan population. Finally, the Spanish conquest in 1531 of Tupac Amaru, the Inca king in the 16th century, brought an end to the Inca Empire.

Incan architecture

The Inca Empire, which reached its height in the middle of the 15th century, is renowned for its exquisite masonry among other things. The workmanship was so refined that finely-cut stones were made to fit tightly without the use of mortar. This street view of Cuzco shows the original Inca masonry. Examples can also be seen at Machu Picchu, a city built by the Incans in 1450, about 50 miles (80 km) from Cuzco, at a height of almost 8,000 ft (2,438 m) on a mountain ridge. The city was abandoned after the Spanish conquest of Inca, and was rediscovered by Harry Bingham in 1911.

Quipu

The quipu was a device comprising knotted ropes, used by Incans to keep records. The extraordinary device was used to record agricultural produce, administrative data and history, among other things. In the sample shown, the quipu is attached to a wooden frame. (Peru; 1430–1532; Museum fur Volkerkunde, Berlin)

distribution of food over long distances as well as storing food in warehouses. There was no system of writing but there was a remarkably innovative way of keeping records. The quipu, made of knotted ropes, was the extraordinary device for

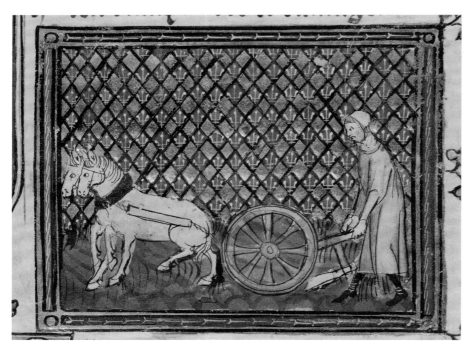

A depiction of the wheeled plow

The wheeled plow was an important invention that increased the agricultural productivity around the 10th century. Earlier the plows merely scratched the surface of the soil; the wheeled plow had a heavy knife that dug under the surface of the soil, resulting in a better seedbed. In this detail from *The Man Obliged to Work for a Living* by Chretien Legouais, one of the earliest wheeled plows can be seen. (France; Medieval; vellum; Bibliotheque Municipale, Rouen)

DEVELOPMENTS IN EUROPE

The new millennium saw many changes in Europe too. Western Europe, which had suffered chaos and the rise of many small kingdoms after the breakup of the Roman Empire, had a brief period of unification during the eighth century under Charlemagne. But this did not last very long, and for the next two centuries, Europe was once again faced with challenges from the north and from the Atlantic.

However, between the 11th and the 13th century, Europe once again saw a revival. The invasions from the north ceased and there was a consolidation of power in the south, particularly in the Castellan region. The period saw a huge increase in the population which was supported by an increased agricultural productivity.

There were many changes in agriculture during this time. Firstly, the three-field system or a system of crop rotation became more common, leading to higher yields. The open-field system, which was adopted during the time of Charlemagne, allowed the farmers to pool their plowing resources, including animals. The plows themselves underwent changes. The wheeled plow with a moldboard—almost certainly inspired by the Chinese design—came into use. This plow allowed the farmer to turn a true furrow and thus make a better seedbed.

In terms of draught animals, the horse collar invented by the Chinese replaced the older harnesses and allowed the animal to breathe easily, thereby giving more traction. This innovation proved to be of great significance for both agriculture and transportation since it allowed the horse to do heavier work.

The agricultural tools being used also underwent some changes, resulting in more efficient designs. The new design of the axe, for instance, allowed a much more efficient clearing of forest land. The norias or water wheels were used for irrigation and windmills for grinding of grain. In parts of northern Europe, land was reclaimed from the marshes and used for agriculture and habitation. New farmland was settled by the German, Dutch and French settlers.

The Islamic influence in Spain and other parts of southern Europe saw the introduction of new crops as well as livestock. Citrus fruits, sugarcane, rice and cotton were introduced and soon there were flourishing cotton and textile manufacturing centers in Venice and Germany. The Moors also introduced a new breed of sheep called the merino in Spain, making Spanish wool a prized commodity in Europe.

The growth in population also led to the growth of urban centers all over Europe. These included the city states of Venice, Genoa and Paris. Several trading centers also developed in northern Europe and on the Baltic coast. Guilds of artisans and shopkeepers were formed, and these played a vital role in the development of trade and industry.

The Impact of the Crusades

The Crusades, which were launched toward the end of the 11th century and lasted till the end of the 13th century, had an enormous impact on the history of inventions in Europe. These were Holy Wars launched with the sanction of the Church with the purpose of "liberating" Jerusalem from the Muslims. The Crusaders were mostly people who had fought the invaders and, with the recession of their threat, were engaged in internecine fighting. The First Crusade was meant to protect the Byzantine Empire from the Muslim threats into its territory, though Christian warriors had already managed to wrest control of Moorish controlled regions like Toledo before that.

The Crusades had a huge influence on Europe. The consolidation of the power of the Pope was one of the consequences of the Holy Wars. But more importantly, the Crusades allowed the transmission of Islamic knowledge to the West on an unprecedented scale.

The Islamic world had an interaction with civilizations in China and India and had learnt their technologies. As the Islamic scientists had translated many of the works of the Greek philosophers into Arabic, these and the several Chinese and Indian texts were now available to the Europeans.

Apart from this, the impact of Islamic architecture was also felt in Europe over the next few centuries. In military technology, there was much which the Europeans learnt from the Islamic world. The wooden fortresses were replaced with stone structures and better siege tactics were also learnt. The introduction of gunpowder and cannons in Europe were to prove crucial in the coming centuries.

In the field of mathematics, the decimal numbering system and algebra—both invented by Indian mathematicians—were introduced in Europe after the Crusades. In medicine, a wide knowledge, not only of the ancient Greek systems but also of indigenous Islamic medicine systems and the Indian and Chinese traditions, was made available to the Europeans.

The spinning wheel was introduced in the 13th century, possibly from India, and this aided in the development of the yarn and textile manufacture. The compass came to Europe from the Islamic world which had obtained it from the Chinese. The impact of the compass, better quality astrolabes and the sextant in navigation proved to be important in the coming centuries when the European navigators and merchants colonized vast areas of the globe.

In the 14th century, the first mechanical clocks were invented in Europe, employing falling weights and the escapement mechanism which produced a cyclic motion. Though the escapement mechanism had been invented much earlier, it was not applied to clocks so far. First appearing in the towers of large Italian cities, the mechanical clocks were one-handed and indicated only quarter hours. They replaced the water clocks which had been in existence for more than a millennium.

The introduction of the Greek works of natural philosophy led to the evolution of Scholasticism between 1100 to 1500, which was essentially an attempt to amalgamate medieval Christian theology with ancient Greek philosophy.

Roger Bacon, a 13th century Franciscan friar, was a great advocate of empiricism and is usually called the founder of the scientific method which places great emphasis on experimentation and observations. His famous work *Opus Maius*—an encyclopedia of all fields of science—has descriptions of various optical phenomena, talks about the process of making gunpowder, discusses the motion of the celestial bodies, and anticipates later inventions such as microscopes, telescopes, spectacles, flying machines, hydraulics and steam ships. In fact, Bacon is credited with advocating the use of magnifying glasses to improve vision. This method had possibly been in use in China as early as the 10th century. In a few years, most of the elite of Europe were using eyeglasses.

Thomas Aquinas, a Roman Catholic priest, was another influential theologian who tried to reconcile Aristotelian ideas with theology. Based on Aristotle's ideas, he advocated his own theories of sense perceptions and intellectual ideas.

The Scholastics played a very influential role in the intellectual revival of Europe in the 13th and 14th centuries. The stress on learning, the growing use of scientific method, and most importantly, the establishment of the Universities—the seats of learning and research—were all, in a large measure, due to them.

The Universities established in this period played a central role in the development of knowledge in Europe in the coming centuries and were responsible for most of the scientific and technological advances in Europe subsequently.

A mechanical clock

In the 14th century, mechanical clocks, weight-driven and regulated by a verge-and-foliot escapement, replaced water clocks. The escapement mechanism was used in clocks for over 300 years. This is a reconstruction of a mechanical clock designed by Leonardo da Vinci. (Italy; Museo Ideale di Leonardo da Vinci, Vinci)

Roger Bacon

The 13th century Franciscan monk Roger Bacon was one of the first proponents of the experimental method in the West. A philosopher and scientist, Bacon studied astronomy, alchemy, mathematics among other disciplines. He described the process of making gunpowder in detail in one of his works.

UNIVERSITIES AS SEATS OF LEARNING

The intellectual situation in Europe in the centuries preceding the 11th century was one of relative stagnation. The smaller kingdoms which came into being after the end of the Roman Empire were never quite at peace among themselves to be able to induce an atmosphere where learning could flourish. While the intellectual landscape was not very fertile in those times, there was one institution which flourished and managed to increase its hold on Europe: the Church.

The intellectual tradition was kept alive by the Church through monasteries. The monasteries had acquired immense power and wealth in the preceding centuries because of the tradition of nobles gifting large estates to them in return for conducting masses. These estates provided a regular income to sustain the monks and nuns in the monasteries

Amalric of Bena at the Old Sorbonne

The University of Paris, also called La Sorbonne after Robert de Sorbon who founded the collegiate institution in 1257, was established in the latter half of the 12th century. It was reorganized in 1970 as 13 autonomous universities. Amalric of Bena taught theology and philosophy at the Sorbonne in the late 12th century. In 1204, the University of Paris condemned the doctrines of Amalric of Bena as heretical. (France; *Grandes Chroniques de France*; Bibliotheque Nationale, Paris)

who could thereby concentrate on the pursuit of knowledge—either theological or philosophical. In fact, the monasteries became the most important centers of learning.

One of the most important functions of the monasteries was to copy and thereby preserve manuscripts, allowing rare documents to be retained for later use. In addition, some of the monasteries had very productive pharmacies since the practice of medicine was not yet a separate vocation. The monks working in these pharmacies maintained herbal gardens and experimented with many new plants. Since wine played such an important role in the Christian tradition, several monasteries experimented with wine-making and grape cultivation, leading for instance to the discovery of Champagne and Benedictine liquors.

In the 11th century, this established order started to change primarily because of the rise of other traditions like the Franciscans who advocated public discourse rather than cloistered learning. However, despite their decreased power, monasteries continued to play an important role in the intellectual life of Europe and monastic life inspired many later thinkers and philosophers.

Within the Church, the emphasis had been on the study of liturgy but this changed in the 12th century when there was a growing need for professional education of the clergy and a more detailed study of the Canon law. The reforms initiated in the Church by Pope Gregory VII around the end of the 11th century were a major impetus for the study of the Canon law. Also, with the rise of the Scholastic movement—a philosophy based on the idea that reason and faith are compatible—there was a growing confidence in rationalism as a relevant method of enquiry into the problems of the world as well as on various philosophical issues.

To facilitate the study of the Canon law and secular disciplines, cathedral schools were established in many small towns to train the clergy. Soon, these migrated to the big cities of the time, most notably Paris and Bologna.

The school at Paris is considered to be the precursor of the university, though chronologically, the University at Bologna was established earlier in 1088. Actually, the first European university was established in the ninth century in Constantinople as a secular institute of higher education, to support the state administration. However, the structure of the university as we see it today has its roots in the later institutions like the Universities at Paris and Bologna.

Though Europe got its first universities in the ninth century AD, these were already in existence in India and China many centuries earlier. The Chinese had academies of higher learning where the principles of Confucianism and classics were routinely taught. In India, the University at Nalanda was established in the fifth century AD though it had been associated with Buddhism for several centuries prior to that. Nalanda was a residential university with over 10,000 students and a couple of thousand teachers. The students stayed in dormitories and the campus had many temples, meditation rooms and ponds. There was a huge library which housed thousands of manuscripts from all over the world. The students at Nalanda were from India and even from China, Korea and Turkey. The university became a major seat of learning, especially of Buddhism, and the famous seventh century Chinese traveler Xuanzang has provided detailed descriptions of the institution, including its administration, funding and courses of study. The university was destroyed in the 12th century by Muslim invaders.

Ruins of Nalanda University

One of the oldest institutions of learning was the Buddhist University at Nalanda, about 62 miles (100 km) from the city of Patna in eastern India. Though the place had been associated with learning earlier, the university was built in the fifth century. It was a huge campus which at its largest had more than 10,000 students who lived in dormitories. The Chinese traveler Xuanzang has left detailed descriptions of the university and its working. The University was destroyed by Muslim raiders in 1193.

By the 12th century, the universities in Europe were well established. They were either funded directly by the students, who paid the teachers, or by the Church or, in some later instances, by the state or the Crown. There were no fixed campuses; classes were held in churches and in the homes of the teachers. Initially, each university acquired a reputation for one or a few disciplines—the University of Bologna, where lawyers were a powerful group, was well known for law, while the University of Paris was recognized for theology.

The students took about six years to complete the first course and were given instruction in the liberal arts like logic, grammar and music. After the bachelor's course, higher studies were usually in the faculties of theology, law and medicine. Women were not allowed in the universities since the students were considered clerics as per the Church laws.

By the 12th and 13th centuries, most universities became organized bodies with their own set of statutes, own courses of studies and so on. A large number of offices of the Church were by now held by people who were masters in the universities. The establishment of universities as seats of learning and scholarship was further bolstered by the conferment on the most prestigious of them of the title of Studium Generale. This honor was bestowed upon the universities by way of a papal bull and the scholars from these places were supposed to travel to other universities and deliver lectures. Apart from the Universities of Bologna and Paris, the newly established Oxford and Cambridge universities in England—founded in 11th and 13th century respectively—were also designated as Studium Generale by the end of the 13th century.

The separation of the seat of learning from the Church had a huge impact on the intellectual revival in Europe. Theology continued to be the course of choice in most universities but now the scholars had more freedom to debate and question. The high points of theological scholarship came with the works of Thomas Aquinas and St Bonaventure.

By the end of the 14th century, universities had spread to northern Europe, Scotland and the eastern part of Europe. In the three centuries, they had increasingly become autonomous of the Church and were now asserting their authority on matters of scholarship. Though there was a crisis of confidence in the 15th century with regard to the impact of scholarship in the traditional mold on addressing issues, the universities continued being the most important places of intellectual work in Europe.

The evolution of universities into their present forms in later centuries was a result of many factors like Reformation and Enlightenment. They played an important role in the Renaissance that followed the medieval period in Europe.

A library in Oxford University

Founded in the 11th century, Oxford University is the oldest university in the English-speaking world. In 1209, some scholars broke away from Oxford University and started the University of Cambridge. With a long history of rivalry, today both are among the most prestigious universities of the world and are jointly referred to as Oxbridge.

EARLY MODERN

DISCOVERY AND EXPLORATION (1400–1700)

RENAISSANCE

The High Middle Ages in Europe in the 13th and 14th centuries had set the stage for a period of intense change in the 15th century. This change, called Renaissance or rebirth, was characterized by a reawakening of interest in the classical learning and scholarship.

The next 300 years saw many profound changes in the economy, polity and the intellectual landscape of Europe. The political and economic center of power that had moved east in the previous centuries moved back to Europe. The revival of Europe after centuries of stagnation was to continue beyond the Renaissance period.

This was the period of exploration of new geographies (the Americas were "discovered" and colonized by the Europeans), of the emergence of new techniques and instruments to study the heavens, of inventions or innovations in technology and also a decline in the feudal structure and growth of commerce. There were also many changes in the religious configuration of Europe that saw the rise of Protestantism and the reaction of the Catholic Church to these events.

One fundamental difference between the Renaissance period and the previous centuries was that the intellectual leadership was taken over by secular people. In the period from 12th to 14th centuries, it was the clerics who had dominated the scholarly activities and this, as seen, had led to the Scholastic philosophy. The 15th century saw the rise of Humanism, an intellectual movement (characterized by the rediscovery and study of the Greek and Roman cultures) which started in Italy. The movement and the revival it brought about spread all over Europe in the next century, in part because of the invention of printing and spread of texts.

The Humanists played an important role in helping Europe free itself from religious orthodoxy and to gain new confidence in human endeavor in the field of ideas and technology. Since the Humanists were not clerics but professionals like lawyers and notaries, they emphasized engagement with society rather than monastic life and stressed the unique capabilities of humans. They advocated a reading of the ancient texts and a systematic enquiry into the physical world in order to understand it and master it.

Map of Sao Jorge da Mina

In the quest to find a new sea route to the rich Cathay, the Genovan navigator and explorer Christopher Columbus led four trans-Atlantic voyages. Columbus' expeditions were financed by the Spanish queen and a consortium of Italian bankers. He landed in what is now the West Indies in 1492 on his first voyage. His subsequent trips were aimed at the conversion of pagans to Christianity and the colonization of the New World. Sao Jorge de Mina, on the Portuguese Gold Coast, was visited by him on his fourth and final voyage.

1403: The Capital of the Ming dynasty moved from Nanjing to Beijing.

1429: Joan of Arc ends the Siege of Orleans.

1453: Fall of Constantinople and the Byzantine Empire.

1455: The Gutenberg Bible printed.

1469: Marriage of Ferdinand and Isabella of Spain creates a united Spanish empire.

1492: Christopher Columbus lands in the New World in his search for the sea route to the East. He founds the first Spanish colony in Hispaniola.

1498: Vasco da Gama discovers the sea route to India via the Cape of Good Hope.

1500: The Incas complete the construction of Machu Pichu.

1513: The Portuguese land in Macau during the Ming Dynasty.

1515–1518: The Ottomans capture Anatolia, Egypt and Arabia.

1517: Luther posts his 95 theses in Germany, beginning the Reformation.

1520: The Spanish conquistador Herman Cortes lands in Mexico. Over the next 60 years, the Spaniards conquered all of Mesoamerica, destroying all the native cultures.

1519–22: A Spanish expedition, led by the Portuguese Ferdinand Magellan, circumnavigates the world, traveling west from Europe, around the southern tip of South America, across the Pacific.

Old sextant on display at the Arsenale in Venice

1526: Start of the Mughal rule in India with the victory of the Mughal emperor Babur over the Lodhis.

1531: The Church of England breaks away from the Roman Catholic Church. The first stock exchange at Antwerp is founded.

1532: Pizarro leads the conquest of the Inca kingdom. After fighting for over four decades, Francisco Toledo finally executed the last of the Incas, Tupac Amaru, in 1572.

1543: The Polish astronomer Nicolaus Copernicus publishes his heliocentric theory of the universe.

1556: Akbar, the Great Mughal Emperor, is crowned King of India.

1582: The first map made with the Mercator projection.

1582: Pope Gregory XIII reforms the Julian calendar by replacing it with the more accurate Gregorian calendar. It is adopted in several European countries.

1589: Galileo Galilei propounds his law of falling bodies which contradicts Aristotle's ideas about gravity.

1600: Potatoes introduced in Europe from the Andes.

1608: The first telescope is built. Galilei improves the telescope to observe the heavens.

1612: The English establish their first factory in India at Surat.

1613: Start of the Romanov dynasty in Russia.

1614: John Napier publishes the first logarithm tables as an aid to calculations.

1620: The *Mayflower* sets sail for North America and the Puritans establish a colony at Plymouth.

1628: William Harvey explains the circulation of blood in the body.

1637: René Descartes invents analytical geometry.

1643: Evangelista Torricelli uses the barometer to measure atmospheric pressure.

1652: Dutch colony established at the Cape of Good Hope in South Africa.

1656: Christiaan Huygens patents the pendulum clock.

1665: Robert Hooke uses a crude microscope to discover cells.

1687: Isaac Newton discovers calculus and the laws of gravitation.

1690: The British establish a trading post at Calcutta.

The study of the classics of Greek and Roman philosophy in the original and in translation was encouraged and this got a major impetus when many scholars migrated to Italy in the middle of the 15th century after the fall of Constantinople to the Ottomans. The learning of the classical Greek and Latin texts gained ground as did the study of secular subjects like history, literature and politics. The ancient Greek texts had been preserved in the monasteries and were also available in the Islamic world. Further, the defeat of the Moors in the Iberian Peninsula made available thousands of works of ancient Greek philosophers to the European scholars.

The Humanist ideas had a profound impact on religion in Europe. This was the time of turmoil in the Church with the Western Schism when many rivals laid claims to the papacy. This unrest was to an extent resolved in 1411 by the Council of Constance and the rise of reformist movements. However, the corruption (sale of indulgences) and the nepotism in the Church went unabated and several clerics, influenced by Humanist thought, attempted a criticism and reform of the Church. These efforts culminated in the work of Martin Luther in 1517, which ultimately led to the Reformation and a break in the hegemony of the Roman Catholic Church in Europe.

In the economic sphere, this was the period which saw spectacular growth all across Europe. First, there was a shift toward increased urbanism, causing a spurt in the population in the cities. There was a substantial investment in agriculture to feed the growing urban population and more commercial crops started being grown. The innovations in technology led to a remarkable increase in manufacturing as the demand for goods increased, especially for military related equipment. The many wars for domination ensured that the demand for cloth, armor, weapons and ships boomed during this period. In addition, innovations in mining technology and metallurgy opened up new mines of iron and copper in various parts of Europe.

The discovery of the Americas also changed the economy of Europe in a significant way. The voyage of Christopher Columbus, which was sponsored by the Spanish emperor, initially did not result in any economic gains. However, in the next few years, especially with the conquest of the Aztec and the Inca Empire, fabulously rich silver mines came under the control of the Spanish crown. The wealth from these mines, and later from the American plantations, contributed to financing not only the opulence in Europe but also the many wars which were fought.

Printing technology

The development of printing was an extremely significant milestone in the 15th century. The impact of printing in the dissemination of Humanist ideas and in the spread of the Renaissance to all parts of Europe cannot be overestimated. Paper-making techniques had already been transmitted to Europe from China and the Middle East by travelers and crusaders in the previous centuries. As a result, several major paper-making centers had emerged in Italy, Germany and France.

Woodblock printing had emerged in China and its use in Europe started in the end of the 14th century, primarily in reproducing the ornamental capital letters on manuscripts. The woodcuts were also used to make religious pictures and cards in large numbers from the early part of 15th century. Subsequently, as the engravers became more adept, small books with both text and images came into circulation. These were mostly religious tracts, though books like compendiums of Latin grammar were also printed.

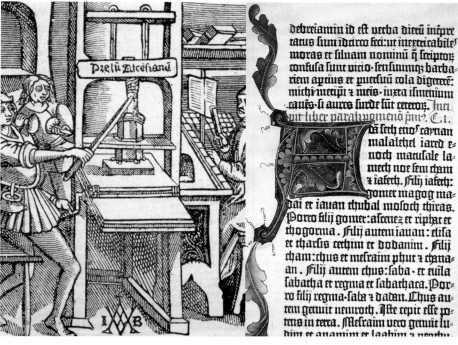

The next step, that of cutting individual letters in wood and then setting them to make the text, was also attempted but with very limited success. The reason for this was that the Roman alphabet letters were much smaller in size than the Chinese characters and hence making them was a much more delicate operation. The blocks made in this way were also very fragile and thus had a limited use in printing.

The transition in the use of wood to metal in printing possibly took place in Korea in the 13th century, though not much is known about it. In

Printing Press of 1498 and Gutenberg Bible

The German craftsman Johannes Gutenberg invented the movable type printing in the middle of the 15th century. His printing press was derived from the press being used in wine-making and paper-making. It had a wooden screw which was operated with a long handle—the operation involved pressing the screw on the paper laid over the type. This design essentially continued for over two centuries. (Engraving from a book printed in 1498)
The first book printed by Gutenberg using the movable type technique was the 42-line Bible in 1455. The three-volume Latin masterpiece used the Gothic type and had no title page or page numbers. It is not known how many copies of the book were printed but 40 copies are still in existence. The decorated initial "A" from the Gutenberg Bible is shown here. (Germany; 1455; parchment; Universitatsbibliothek, Gottingen)

Europe, the first steps were taken by metal founders and die cutters who made individual letter dies and then used molten lead to form a plate. This process was much faster and durable than using wood but there was a problem of alignment in the plates.

The stage was thus set for the emergence of the true movable type printing and the printing press. Johannes Gutenberg is credited with this innovation in the middle of the 15th century. Gutenberg, a resident of Mainz in Germany, had worked with goldsmiths and was thus familiar with metal working. In the process he devised, molten lead was poured to make a matrix and a mold into which an alloy of tin, lead and antimony was poured to make the type. The alloy was crucial since it was much more durable than just lead.

The other innovation in printing was the development of the printing press. The original press was possibly inspired by the design of the agricultural press used for centuries. It consisted of a flat, fixed lower surface on which the typeset was placed and a movable upper part which could be screwed tight with the lower surface. The text to be printed was first composed or typeset in a wooden strip. This was then locked or screwed into a metallic frame and ink was applied to it.

A major innovation in inks also helped the spread of printing. Hitherto, water-based inks had been used for texts. Flemish painters around this time were experimenting with oil-based paints and perhaps this was the inspiration for developing the oil-based ink for printers, which was longer-lasting than water-based inks. Once the composed text had been inked, paper was placed on it and the two sides were pressed in the vise to get an impression of the text on the paper.

The results from the printing press were far superior to those obtained by the woodblock brushing techniques—the printing was much sharper and one could print on both sides of the paper. Gutenberg used his printing press to print 180 copies of a 42-line Bible in 1455 and there was no looking back from there.

Printing rapidly spread to other countries from Germany, and in a few decades, all major cities of Europe had a flourishing printing industry. The importance of printing in the dissemination of Renaissance and Humanist ideas was enormous. Knowledge, whether in terms of new books or ancient texts or even religious texts, was no longer the preserve of the wealthy and the clergy who alone had access to the hand-copied manuscripts. It was now possible to efficiently and quickly produce many copies of books and pamphlets at a low cost.

Art

The Renaissance in Italy in the 15th century was responsible for many revolutionary ideas and innovations but is possibly best known for its art. One of the changes in painting that came about during this time was the introduction of perspective in painting. Although earlier artists had used this technique, it was only in the 14th century that this gained popularity as an artistic style. It was around the same time that Realism in art became popular. The artists attempted to portray objects as the eye perceives them, with the complex interplay of light and shadows, the nuances of perspective and the details of anatomy.

Among the most famous painters from this period is Leonardo da Vinci. Leonardo, who has been variously described as a genius, a complete Renaissance man and a polymath, made enormous contributions to a variety of fields. In painting, he studied human anatomy in great detail by

The Adoration of the Magi

In the early 15th century, the mathematical laws of perspective were discovered by the Italian architect Filippo Brunelleschi. Some of the ideas were already known to the Greeks but these were not implemented in art. Renaissance painters quickly adopted these principles and used them in their work. For the first time, paintings seemed to be depicting objects as we see them, with a feel of spatial depth. This is the background perspective sketch from Leonardo da Vinci's *The Adoration of the Magi*. (Italy; 1481; pen and ink on paper; Galleria degli Uffizi, Florence)

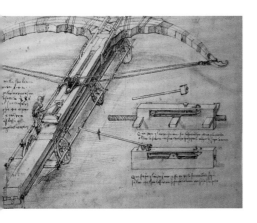

issecting corpses. He even dissected animals o understand their anatomy. He made many aintings and experimented with varied techniques nd styles.

Leonardo was not just a painter—he also a very een observer of nature, a scientist and an engineer. lis journals and notebooks contain innumerable rawings and ideas of technologies which were nuch ahead of the times. Leonardo's ideas not only eflected an engineering inventiveness but also a eep understanding of the science of mechanics. le designed machines with a differential gear ystem, a flying machine (which was inspired by is studies of bird flight), a moving fortress (or a ank) and a helicopter. He also studied whirlpools nd eddies and made detailed drawings which eflect an understanding of the underlying physics f the phenomenon.

Apart from being an accomplished painter nd artist, Leonardo was also an architect and a nilitary engineer. Although he did not erect any uildings, he spent all his life designing buildings f various kinds. He wrote copiously on various ssues in architecture and urban planning. Some f these works had a tremendous influence on he practising architects at that time and were nfluential in the development of the architectural tyle that has been associated with this period.

A contemporary and in some senses rival of eonardo da Vinci, Michelangelo Buonarroti vas also a polymath. An extremely accomplished ainter and sculptor, he had an enormous influence n artists during his lifetime and afterward. His vorks like the frescoes on the ceiling of the Sistine Chapel, *David* and *Pieta* continue to be among the est examples of art of that time.

Astronomy

n the middle of the 16th century, a Polish stronomer changed the way people perceived the eavenly bodies. Nicolaus Copernicus was on his deathbed when his magnum opus, *De revolutionibus orbium coelestium libri VI* (*Six Books Concerning the Revolutions of the Heavenly Orbs*), was ublished. This book, which has been called one

of the most influential books in the history of science, changed not only the field of astronomy but also led to a revolution in the worldview of human beings.

For centuries, astronomy and astrology had been seen as common disciplines with the study of the heavens being done with the purpose of "predicting" future events. The dominant theory of the universe in the Western world was the Aristotelian view which postulated that the earth was at the center of the universe and the planets are fixed on material and yet invisible spheres, forming a series of concentric circles. This view, which had held for almost two millennia, was deficient in explaining the observed properties of the planets. For instance, planets were known to vary in their brightness and it was hard to explain how this could be understood if they were moving at a fixed

distance from the earth. The planets were also known to have occasional retrograde motion which was inexplicable in the Aristotelian framework. Claudius Ptolemy, in the first century AD, had tried to rectify this with his theory of deferents and epicycles. This enormously complicated theory was based on a geocentric model.

Copernicus proposed a radically different approach to the problem. He postulated that the sun, instead of the earth, is at the center of the solar system. The earth and all the other known planets revolve around the stationary sun in well-defined orbits at various distances from the sun. This theory, he advocated, would solve the problem of varying brightness as well as the retrograde motion of the planets. In doing so though, it

Leonardo da Vinci's Giant Crossbow

Painter, engineer, sculptor and architect Leanardo da Vinci was possibly the most famous of Renaissance figures. His *Mona Lisa* and *The Last Supper* are arguably the most famous works of art. He was a rare genius whose thirst for knowledge, scientific enquiry and technical inventiveness was much ahead of the times. His notebooks contains drawings of mechanical devices which were not invented till much later, such as the airplane, the helicopter (and other flying machines), the parachute, the submarine, the armored car, the ballista (a giant crossbow), rapid-fire guns, the centrifugal pump, ball bearings and the worm gear. This drawing of a giant crossbow was made in 1499. This gigantic weapon was conceptualized as having two firing mechanisms—one involving the release of a holding pin by hitting it with a mallet.

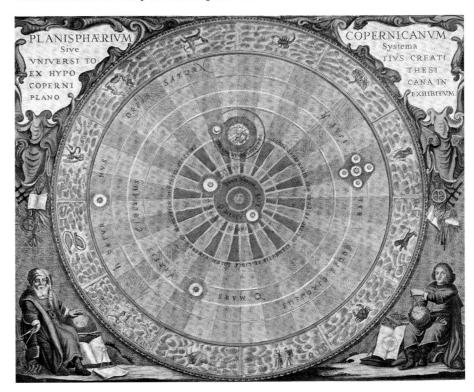

The Copernican system

Nicolaus Copernicus was a Polish astronomer who first proposed the heliocentric theory in 1543. Copernicus's theory, which was more successful in explaining the planetary phenomenon than the earlier Aristotelian or Ptolemaic theories, was very influential in shaping the thinking of the founders of the scientific revolution like Galilei, Newton and Descartes. In fact, the idea of moving the center of the universe from the earth to the sun had profound implications for the dominant worldview. This is a 17th century representation of the Copernican world system.

Galileo Galilei

The Italian scientist Galilei can be called the father of modern science. A firm believer in the experimental method and the use of mathematics in formulating physical laws, he conducted many experiments in a variety of fields. His use of the telescope to observe the heavens affirmed his faith in the Copernican theory which led to his conflict with the Church. His work in understanding the laws of motion preceded the formulation of these laws by Newton.

Mercator world map

Gerardus Mercator, a 16th century Flemish cartographer, devised a mathematical way of projecting the map of the globe on a flat surface. As the Mercator projection maps represented lines of constant true bearing, they became the standard for navigational purposes. Mercator also coined the word Atlas for a collection of maps. This image shows the "Map of the World", from the *Atlas sive cosmographicae*. (Duisberg, Germany; 1585)

would have to abandon the edifice of all of natural philosophy as had been known for centuries in the Western world, not to speak of the conflict with the position of the all-powerful Catholic Church.

The conflict with the Church came to a head with the work of another scientist, Galileo Galilei. Galilei, an astronomer and mathematician, was a major figure in this period. His contributions to physics and astronomy and, more importantly, the method of studying nature were fundamental to a changed worldview of nature. Galilei formulated the laws of falling bodies by conducting his now famous experiments of dropping weights from on top of the Leaning Tower of Pisa. This and his other discoveries on the nature of motion and inertia were all based on experimentation. Galilei's ideas on motion were in contradiction with the reigning Aristotelian notions about nature. This made him unpopular among his peers and is said to have been responsible for the cancellation of his contract at the University of Pisa.

The major contribution of Galilei to science was not only the specific laws of motion and falling bodies that he formulated but, more importantly, the methodology. Galilei was among the first to move away from a qualitative description of natural phenomenon to a mathematical one. He was also an avid experimentalist who propagated that the touchstone of the correctness of any theory was its experimental verification.

But it was Galilei's work in astronomy which brought him the most fame. Galilei improved upon the telescope which had been invented a few years earlier in the Netherlands to make a very useful device to view far-off objects. This was an extremely practical device for the merchants in Venice since they could get advance information on the arrival of cargo ships into the harbor.

Galilei used the telescope to systematically study the heavens and discovered many new objects like

the moons of Jupiter and the unusual appearance of Saturn. As a result of his observations, Galilei was convinced that the heliocentric view of the heavens as propounded by Copernicus was the correct one. This marked a major landmark in the beginning of the scientific revolution. His views on the heavens were considered to be heresy and Galilei had to face the wrath of the Church. While it is not clear how the controversy ended—whether Galilei rescinded his ideas or whether the Church allowed him to get away with just an admonition—his work gave a new direction to the scientific community.

Galilei made several contributions to technology apart from perfecting the telescope. He built the first recorded compound microscope to see extremely small objects like details of insects. He also studied the motion of pendulums which was very useful in the making of the pendulum clock some years later. He invented an air thermometer which used the expansion of air to measure temperature.

Navigation

The 16th century was the century of exploration. Navigation was important for the explorers as was the technology for making better and more rugged ships. In 1568, the Flemish cartographer Gerardus Mercator introduced a technique of making maps which was immensely useful for the navigators. In 1569, the geographer introduced a projection in which the lines of the meridian were parallel as were the lines of latitude. The meridians were equally spaced while the horizontal lines of latitude get further and further as we move away from the equator. The advantage of this projection or way of depicting the spherical earth on a flat surface was that it allowed the navigators to chart a straight-line course on the map. Mercator also introduced the term atlas for a collection of maps.

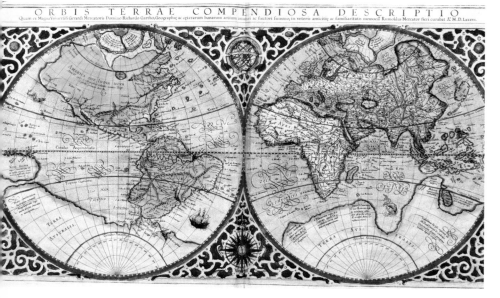

ORBIS TERRAE COMPENDIOSA DESCRIPTIO

THE DEVELOPMENT OF SHIPS AND NAUTICS

The 15th century saw the beginning of the age of exploration in Europe. Trade and exchange of goods had been going on for centuries with Asia and Africa in the previous centuries. Most of this trade (with the exception of the Mediterranean trade) was by overland routes, primarily the Silk Road which connected China to Europe through central Asia, with offshoots going off to Iran and Afghanistan. With the rise of the Ottoman Empire in the middle of the 15th century, these trade routes were blocked. As a result, merchants were forced to look for alternative means of transporting goods to satisfy the demand of a growing population in Europe for Asian goods.

As already seen, river and ocean transportation has a long history—from the simple reed boats used in Mesopotamia and Egypt to the many oared vessels used by the Romans and the Greeks. With the increase in the distance that needed to be traveled for trade, ships had to have more propulsive power. This could only be done by increasing the surface of the sails available—either by making the sails bigger or having more of them, or both. Prior to this time, oared vessels which traveled close to the coast without any need for directional navigation in the open seas were enough to carry out short distance trade.

The search for a sea route to the East Indies had led Christopher Columbus to the Americas. On August 2, 1492, Columbus and his crew sailed westward from Spain on the three ships *Santa Maria*, *Nina* and *Pina*. Columbus made three more voyages to America after this—between 1493 and 1504. Besides discovering the New World for the Europeans, the highpoints of these expeditions were the colonization of Hispaniola and the discovery of the South American mainland.

In 1497, the Portuguese explorer Vasco da Gama reached Calicut on the western coast of India by going around Africa. Interestingly, in 1500, another Portuguese, Pedro Cabral, was also sailing to Calicut in search of pepper when he reached the coast of Brazil! In a few years, the Portuguese had established trading centers at many places in the East, including India, along the African coast and as far away as Malacca.

The voyages of the Portuguese explorer Ferdinand Magellan opened up new vistas in exploration. In 1519, Magellan sailed with five ships to Brazil and then down the South American coast along Patagonia, to finally find a route westward to the Pacific. After a long and arduous journey across the huge expanse of the Pacific, Magellan reached

the Philippines where he was killed. The remaining ships carried on with the voyage. In 1522, on their arrival in Spain, the world came to know that there was indeed a route to the East via the West! Magellan's circumnavigation was truly a revelation to the Europeans.

The age of discovery and exploration which started in the 15th century was obviously due to the advances in ship-building and navigation. The 15th century saw the emergence of the caravel—a small, maneuverable ship with two or three masts. This was the ship used by the Portuguese, especially for their exploration along the western coast of Africa. The larger ships were suitable for ocean voyages but

could not be used in rivers or in shallow waters. The caravels were ideal for this purpose and hence were extensively used. The 15th and 16th centuries were also the time of the full-rigged ship, which is primarily a ship with three or more masts to hang the sails. More masts meant that the length of the ship had to be elongated to sometimes two to three times the ship's width or beam. The first such ship was developed in Genoa and was called the carrack. These were the first ships which were large enough to carry enough provisions for long journeys and were stable in the rough oceans. They were also capable of carrying guns on board which was a distinct advantage since, unlike earlier centuries, wars were fought on the seas for control of trade. The carrack was the workhorse for most of the Portuguese and Spanish explorations. Columbus' flagship on his maiden voyage, the *Santa Maria*, was a carrack which was accompanied by several caravels.

By the 17th century, almost all the ships being used for long voyages in Europe were fully rigged.

Columbus' fleet

The discovery of the New World by Columbus in 1493 was certainly an event of monumental importance for Europe. The subsequent colonization of the Americas gave the Europeans access to almost unlimited quantities of gold and silver, huge plantations that needed slaves from Africa, and a native heathen population that could be converted. The fleet used by Columbus in his maiden voyage across the Atlantic in 1492 can be seen in these replicas on a tour to celebrate 500 years of Columbus' discovery of America. It comprised *Santa Maria*, the flagship, which was a carrack, and two other caravel-type ships *Nina* and *Pinta*. These ships, specially designed for the purpose of exploration, reflected development in ship-building in the 15th century. (NASA: Kennedy Space Center)

Jacques Cartier's ship

French explorer Jacques Cartier (1491–1557) led three expeditions to Canada, in 1534, 1535, and 1541. Cartier sailed in the Gulf of St Lawrence and paved the way for French exploration of North America. This is a depiction of his flagship, called *Le Grande Hermine*, in *Rarete des Indes sauvages*. The ship was was a galleon—a slender vessel which had evolved from carracks. (France; Pen and ink on paper; Bibliotheque Nationale, Paris)

These were used by explorers like Columbus, Francis Drake, Walter Raleigh and Sebastian Cabot among others to explore most of the world.

The opening up of the sea route to the East had enhanced the prospects of trade immensely. European merchants visiting India and China realized that there were a large number of goods which could be traded apart from the staple of spices and cloth. Factories were set up in the coastal regions and were manned by Europeans to facilitate trade.

On the other hand, the Americas had a huge potential for not only bullion—which was crucial to the Europeans since they needed this to finance the trade deficit with India and China—but also plantation crops. Unlike the East, the Americas had no manufacturing and hence the situation was different. Larger ships to hold more men and goods were needed for the colonies in the Americas.

The trade with the East was hugely profitable and hence there was a severe clash between the various European powers for control of the trade routes. The most powerful nation in the 17th century was Holland with its control of the East Indies spice trade. The Dutch used a kind of ship called the Dutch fluyt which was a long, narrow ship with three masts. It was designed to carry as much cargo as possible in the large hold beneath the deck.

But soon the English gained prominence by the virtue of their trade with India where they were fighting the French and the Portuguese. The English invested heavily in ship-building and the result was the huge, full-rigged East India merchantmen ship which could travel long distances without needing to hit the land. The English enacted law which restricted English trade to English ships. Th provided a major boost to ship-building in England As a result, the total tonnage of English shippin almost doubled in the late 17th century.

The age of discovery and exploration, whic began in the 15th century, changed the map of th world. The opening up of the Americas was followe by a rapid destroying of the existing empires i Mesoamerica and South America. The Aztecs wer conquered in 1521 while the Incas were vanquishe in 1532. The conquistadors used firepower t overcome the disadvantage they had in numbers But by far the most devastating effect was that o the diseases. The natives of the newly discovere continent had no immunity to germs brought by th Europeans and hence were decimated by epidemic of smallpox and other contagious diseases.

The Americas gave Europe new crops like tomatoe (which were propagated throughout the world b the Spanish), capsicum and most importantly, th potatoes, which became a staple crop in large part of Europe soon after the Spanish brought it fro Peru. Another important import from the America was maize which soon became universal.

The advances in technology, especially of ship building and of weapons, ensured that the European controlled the trade routes to the much mor prosperous East. What began as trading between tw parts of the world eventually resulted in military conquest of large parts of Asia and Africa and the establishment of European colonies in the coming century.

Gregorian Calendar

In 1582, Pope Gregory XIII initiated a major reform in the Julian Calendar which had been in use till then. In the Julian Calendar, the year was 365 ¼ days, with leap years every four years to take into account the correspondence of the calendar seasons and actual seasons. However, since the actual year was slightly different from the assumed 365 ¼ days, the calendar accumulated about one day's error per century.

Pope Gregory's reform consisted of restoring the spring equinox which had by now shifted to March 11 instead of March 21, where it had initially been designated in 325 AD. The net effect was to advance the calendar by 10 days in October 1582. According to the Julian calendar, all years divisible by four were leap years. The basic change in the Gregorian calendar followed the rule that all years (except for years divisible by 100) exactly divisible by four are leap years; however, the centurial years exactly divisible by 400 are still leap years. The new calendar was adopted by the countries in Europe over the next couple of centuries and is now taken as the standard.

Science and Mathematics

The 17th century saw a virtual explosion of innovations in many fields—some directly useful while others being of a fundamental nature which enhanced knowledge in areas like medicine and mathematics. This was the century which saw some great mathematicians who worked on a variety of subjects. John Napier, a Scottish mathematician, published the first logarithm tables in 1614. These tables simplified calculations of trigonometric quantities. In France, mathematicians laid the foundations of number theory, analytical geometry and the science of probability. Pierre de Fermat, René Descartes and Blaise Pascal were prolific mathematicians at this time who worked on a number of subjects.

Fermat, a lawyer by profession, is widely regarded as the founder of the modern theory of numbers. Inspired by the work of the ancient Greek mathematicians like Diophantus, Fermat made many fundamental contributions to the number theory. Although he proved many results in this field, he is best-known for his "unproved" result which has since been called Fermat's Last Theorem. This conjecture, which was proven only in 1995, has seen many attempts at proof by some of the best mathematicians.

Descartes, a contemporary of Fermat, was another great philosopher and mathematician. He invented analytic geometry and also laid the foundations of the infinitesimal method, which was later used effectively by Isaac Newton and Gottfried Leibniz to invent calculus. The Cartesian coordinate system, which is ordinarily used in the drawing of graphs, was also invented by him.

Blaise Pascal was another brilliant scientist, mathematician and philosopher. Barely 19, he constructed a mechanical calculator to help his tax collector father with his work. Pascal also laid the foundations of the science of probability, after reportedly being asked by a gambler friend to help him! He worked on hydrostatics as well and discovered an important property of fluids.

By far the most significant scientific figure of the 17th century was Isaac Newton. Newton, quarantined at his farm due to the outbreak of the plague in 1665, worked out the fundamental laws of motion as well as the law of gravitation. His book *Philosophiae Naturalis Principia Mathematica*—hailed among the most influential books of the millennium—contained a detailed exposition of the laws of motion and gravitation, and laid the foundations of the science of mechanics.

Pascaline

Blaise Pascal, a French philosopher, physicist and mathematician, developed a mechanical calculator in 1642 to help his father, a tax commissioner, with arithmetic. The calculator, called Pascaline, could only add and subtract—the numbers were dialed on metal wheels and the solutions appeared in the boxes on top. The device was the second of its kind—the first calculator was invented by the German polymath Wilhelm Schickard in 1623. (France; 1642; Wood and metal; Conservatoire National des Arts et Metiers, Paris)

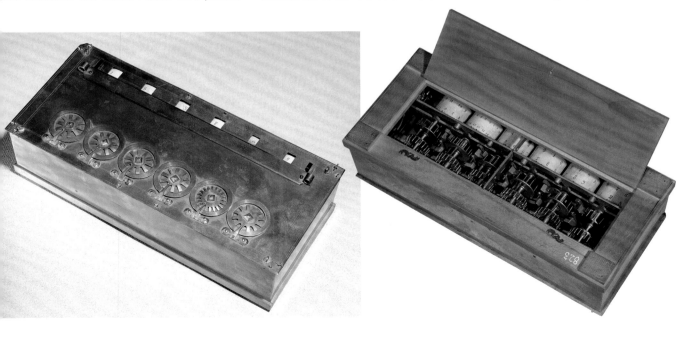

Isaac Newton

Sir Isaac Newton (1642–1727) was the towering figure with whose work the scientific revolution culminated. Newton did path-breaking work in optics, laid the foundations of the science of mechanics and formulated the laws of gravitation. He also, together with Gottfried Leibniz, invented the powerful technique of calculus in mathematics. Newton's magnum opus *Principia* is widely regarded as the most influential work in science of the millennium. (UK; 1702; Oil on canvas; National Portrait Gallery, London)

Optical phenomena of Newton

These diagrams from *Opticks* show Newton's work on light and spectrum. The figures elucidate Newton's theories of dispersion, total internal reflection and refraction. (London; 1704)

The law of gravitation is undoubtedly one of the most significant achievements of the period. Using his theory of gravitation, Newton was able to explain the motion of the planets and the motion on the earth under gravity. Already in 1615, the German astronomer Johannes Kepler had deduced the laws of planetary motion. But it was Newton who provided the theoretical framework to understand the empirically observed Kepler's laws. The importance of Newton's work to modern science was enormous. The use of mathematics to formulate empirically verifiable theories was a great achievement of the *Principia*. He also worked on the theory of optics and invented the reflecting telescope. He propounded a theory of light where he postulated that light was made up of particles or corpuscles and also developed a theory for color.

Meanwhile, a contemporary of Newton, the Dutch mathematician Christiaan Huygens also did some pioneering work in optics—he derived the laws of reflection on the basis of the concept that light is a wave motion transmitted through ether. Huygens was very interested in improving the quality of clocks since they were absolutely critical for reliable navigation. This being the age of exploration, accurate clocks which could be used to find the longitude accurately would have been extremely valuable for navigators and merchants. In 1656, Huygens patented the first clock based on pendulums, which was fairly accurate. In 1675, he patented the first pocket watch and also made improvements in the design of the microscope to include eyepieces which were corrected for chromatic aberration.

Among the other useful devices invented in the century was the Vernier scale which was invented by the French mathematician Pierre Vernier in 1631. This allowed navigators a much more accurate reading with instruments such as the sextant. It was also very useful in surveying since for the first time there was a method to accurately measure very small angles and distances. An enhancement of the Vernier scale was the measuring device called the micrometer which was invented by an

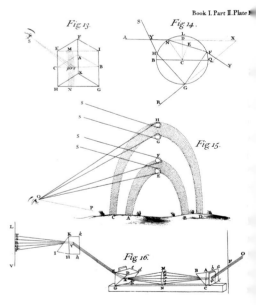

English instrument maker, William Gascoigne. This allowed a far more accurate measurement of angles and was most useful in telescopes.

The measurement of air pressure by a barometer was first achieved by the Italian mathematician Evangelista Torricelli in 1643 by using a tube filled with mercury. While there was no direct use of the barometer, it was an important "enabling" invention which allowed scientists to study the nature of vacuum. In fact, it was in an effort to address the problems of a suction pump that Torricelli first invented the mercury barometer.

Newton's reflecting telescope

Beginning in the mid-1660s, Newton conducted a number of experiments on the composition of light and came up with the modern theory of optics. His experiments on light formed the basis of the invention of the reflecting telescope by him in 1668. The existing telescopes used glass lenses, which caused refraction of light and led to chromatic aberration, a problem in focus. Newton solved this by using mirrors, instead of lenses, in telescopes, which reflected the light to bring it to a focus. A similar instrument was designed by the Scottish mathematician James Gregory in the early 1660s.

Medicine

The impact of experimentation was also felt in the area of medicine. The early Renaissance men like Leonardo da Vinci had already demonstrated the use of dissection in understanding anatomy. William Harvey, an English doctor, was the first person to describe the circulation of blood and the functioning of the heart as a pump. The Spanish doctor Michael Servetus had done this before Harvey but his work never got publicized. Of course, the Islamic scientists had already referred to the pulmonary system several centuries ago.

In the middle of the 17th century, the Dutch trader Antony van Leeuwenhoek used the microscope to observe bacteria and spermatozoa for the first time. A whole new world, invisible to the naked eye, opened up with this—a development which would have a profound impact on the understanding of biology and medicine in the coming centuries. In this, the impact of the microscope was similar to that of the telescope since both of these devices of the 17th century opened up new vistas which had been hitherto invisible to human beings.

By the end of the 17th century, the world was ready to enter the Modern Age. Indeed, the 17th century has been called the Early Modern period of Europe. Modern science was now gaining ground with the works of Galilei, Fermat, Pascal, Descartes and Newton. This gradual replacement of dogma and tradition by scientific reason was one of the key achievements of the 16th and 17th centuries. The development of science was essential for the innovations in technology which led to the Industrial Revolution in the coming centuries.

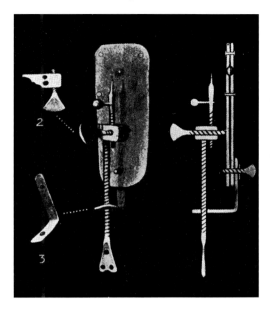

Microscope

Antonie van Leeuwenhoek used to grind lenses and use them to study small objects as a hobby. He made the first simple microscope and used it to observe bacteria and protozoa. His observations were important in refuting the dominant theory of spontaneous generation of life. Even though his microscopes were not very powerful, the lenses were of very high quality. Microscopes were improved upon subsequently and served as important tools for biologists. This is a diagram of one of Leeuwenhoek's microscopes used in the 17th century.

TELESCOPE

Sometime around the beginning of the 17th century, the telescope made its appearance in the Netherlands. It is not very clear as to who the actual inventor of this combination of lenses in a tube was, though it is clear that the Dutch lensmaker Hans Lippershey did contribute to this invention. Within a few years, this contraption had traveled to all parts of Europe. In far away Venice, the Italian scientist Galileo Galilei heard about this wonderful device which could make faraway objects appear near. He immediately realized the importance of this discovery and soon made a much improved version of it which was used by Venetian merchants to locate incoming ships.

Galilei kept on improving the telescope and using it for observing the heavens. With its help, he discovered four of the largest Jovian moons—Io, Ganymede, Europa and Callisto. His observations of the motion of the Jovian satellites around Jupiter played an important role in his abandonment of the geocentric theory and acceptance of the Copernican heliocentric theory. Galilei also observed the phases of Venus, the rings of Saturn and even sunspots with his telescopes.

The telescope was to play an extremely important role in learning about the heavens in the 17th century. Christiaan Huygens used an improved telescope in the middle of the century to observe the rings of Saturn. He also observed that the Orion nebula was not a single heavenly body but a conglomeration of stars.

Early 20x refractor telescope

In 1609, the Italian scientist Galileo Galilei developed the early 20x telescope which could be used for astronomical observations. The telescope had been invented earlier but Galilei improved it to get an upright image. He used it to observe the surface of the moon; he also discovered the moons of Jupiter with it in 1610.

THE MODERN ERA
MANUFACTURING AND TECHNOLOGY (1700–1914)

THE CHANGING FACE OF THE WORLD

The 17th century was a period of many great innovations, especially in the field of science. However, the impact of these new ideas and theories on technology was very limited for a variety of reasons. Although the period saw some of the giants of modern science, like Galilei and Newton, the overall educational system did not promote innovation, creating a huge gap between theoretical ideas and practical applications.

This situation changed dramatically in the 18th and 19th centuries with the introduction of many new technologies. The Industrial Revolution, which started in England in the latter part of the 18th century, changed the production process in a fundamental way.

Apart from the Industrial Revolution, the 18th century also was a period of intense change in the polity and society of Europe, and indeed the world. In 1707, the Acts of Union were passed by the parliaments of England and Scotland to merge the kingdoms of England and Scotland into a single kingdom of Great Britain. On the continent, the beginning of the 18th century saw the Spanish War of Succession which lasted over a decade and resulted in the decrease in French power in Europe. In Prussia, Frederick the Great came to power and ruled for almost half a century. His regime saw a modernization of the civil service as well as an increased religious tolerance. Finally, toward the end of the century, a series of events which culminated in the French Revolution, took place. The French Revolution had a profound impact on the politics not only of Europe but also of the whole world.

Across the world, in the Indian subcontinent, the last of the Great Mughals, Aurangzeb, died in 1707, leading to a weakening of the central

Ford Model T

Ford Motor Company launched the Model T on October 1, 1908, which revolutionized the automobile industry. Hailed as America's Everyman car, it was easy and cheap to own and repair. Ford is credited with modernizing the industry with innovations such as the assembly line and the dealer–franchise system. By 1927, when the production of Model T was finally abandoned, 15 million cars had been produced and the number of Americans owning cars had risen from 8 to 23 million.

1701–1714: Spanish War of Succession. It ended with the Treaty of Utrecht in 1713 which recognized Philip as the King of Spain. It also gave the English the exclusive right of slave trading in the Americas for 30 years.

1707: The Act of Union merging Scotland and England into Great Britain is passed.

1715: The first Jacobite Rebellion.

1722: The rule of the Safavid dynasty in Iran ends after more than two centuries with the sacking of Isfahan by the Afghan armies.

1740: Frederick the Great becomes the monarch of Prussia.

1756–63: The Seven-Year War between various European powers.

1757: The British win the Battle of Plassey and establish their control over Bengal in India.

1762: Rousseau publishes *The Social Contract* which lays down the fundamental philosophy of the rights of citizens. The book is enormously influential in shaping the ideas behind the American and French Revolutions.

1773: The Boston Tea Party. American patriots dump British cargo into the sea as a revolt against British control over trade.

1775–83: The American Revolutionary Wars for independence.

1776: Adam Smith publishes *The Wealth of Nations*, the ideological basis of capitalism.

1780: The Industrial Revolution begins in Britain.

1788: The first European settlement established in Australia.

1789–99: The French Revolution topples the regime of Louis XVI and Marie Antoinette. The revolution unleashed a reign of terror and established a republic for a few years. The Rights of Man and Citizen formed the preamble to the French Constitution. Finally, Napoleon Bonaparte, an army general, was proclaimed emperor in 1804.

1792–1808: Britain, America and France pass laws that ban the trade of slaves.

1801: Great Britain and the Kingdom of Ireland unite to form United Kingdom.

1803: The French sell off their territorial claims in North America with the Louisiana Purchase.

1810–1825: Simon Bolivar wages war against Spain to liberate Latin America from Spanish rule.

1815: Napoleon is defeated in the Battle of Waterloo. Britain emerges as the dominant power in the world.

1840: The first postage stamp, Penny Black, is introduced in Britain.

1842: China, having lost the Opium Wars, is forced to sign the humiliating Treaty of Nanjing, granting enormous concessions to foreigners in trade as well as giving Hong Kong to Britain.

1846–48: The US declares war on Mexico. Mexico is forced to give up almost half its territory.

Penny Black, the first postage stamps, issued in 1840

1848: Karl Marx and Engels write the revolutionary *Communist Manifesto* which calls for the establishment of socialism.

1851: The Great Exhibition at Crystal Palace, showing industrial, economic and technological strengths of Britain.

1854: Japan opens up to foreign trade after Perry demonstrates America's naval prowess.

1857: The sepoys of the East India Company revolt in India. The British

manage to crush the revolt but the rule of East India Company is replaced by the rule of the Crown through a Viceroy.

1858–1869: The Suez Canal is built from the Mediterranean to the Gulf of Suez. It cuts down the travel time between Europe and the East substantially.

1860–61: The Southern States secede from America after Abraham Lincoln is elected the President of the US.

1884: The Prime Meridian, 0 degrees longitude, is established at the Greenwich Observatory.

1889: Japan promulgates the Meiji Constitution, a major step in political reform.

1896: The Olympic Games are revived in Athens, Greece.

1904–05: Russia and Japan fight the Russo–Japanese War for control over the resource-rich Manchuria and Korea.

1905: Tsarists troops massacre workers and trigger the 1905 Russian Revolution.

1910: Four colonies are merged to form the Union of South Africa with political power resting with the white minorities.

1914: Archduke Franz Ferdinand assassinated. World War I begins.

uniting power. This allowed the British East India Company to consolidate its hold in India, especially after the Battle of Plassey in 1757, which marked the beginning of two centuries of British rule in India. Toward the end of the 18th century, a whole new continent, Australia, was colonized as a penal colony by the British. With this, essentially the whole globe, with the exception of Antarctica, had been explored and mapped by the Europeans.

In the New World, the American War of Independence led to the breaking off of the American colonies from England in 1776. The Declaration of Independence, proclaiming the pursuit of Life, Liberty and Happiness, was adopted in the same year. This struggle for self-government and the Declaration of Independence served as an inspiration for political movements across the world.

INDUSTRIAL REVOLUTION

Europe in the 18th century was undergoing many changes. A European identity was being forged, which was assisted by a common language of law and diplomacy—Latin—as well as the common Roman Law. The Renaissance, with its attendant discoveries in science, had opened up the mental horizons of the educated elite in Europe. Printing, which grew rapidly, was responsible for the widespread dissemination of radical ideas among the population throughout Europe.

This was the Age of Enlightenment in Europe— a period when the old, established ideas were being questioned and there was a growing faith in human reason. The intellectual movement, which had its roots in the Renaissance Humanism of the previous century, swept Europe in the late 17th and 18th centuries. The new scientific methods of observation and experimentation were used by scientists like Galilei and Newton to understand and study nature. Nature was increasingly being seen as amenable to reason. The methodology of science and mathematics had an impact on philosophy, and this culminated in France, where the ideas of Descartes were very influential. As Cartesian thought was immensely powerful, rationalism soon became the dominant intellectual trend in Europe. Among the towering figures of Enlightenment were François-Marie de Voltaire, Baron de Montesquieu, Immanuel Kant, John Locke and David Hume in Europe, and Thomas Jefferson and Benjamin Franklin in America.

The economic landscape was also changing rapidly. Trade with the colonies, as well as the fabulous riches of the New World, had contributed significantly to the economic well being of a section of the population. The amassing of huge capital from the colonies, like for instance by the East India Company, played an important role in the investments which were made during the 18th century. However, even with increased wealth, communications were still primitive as was the educational system, which still stressed the study of classical knowledge and did not encourage a critical outlook.

There was already a well-established merchant class which had prospered hugely from trading. Manufacturing took off in a big way, though the dominant mode of production continued to be domestic, with individual households producing goods for the market, but gradually there was a growth in consumerism across Europe.

Agriculture

The Industrial Revolution, which started in Britain in the latter part of the 18th century, brought about an economic change of unprecedented dimensions. The population of Europe grew substantially during the 18th century. The previous century had seen a decrease in population because of the loss of agricultural productivity due to the Little Ice Age. In addition, epidemics of measles, typhoid and tuberculosis took a huge toll on the population which was largely malnourished and living in squalid conditions. Population was also decimated by the Black Death, an epidemic of plague, as well as continued wars being fought in most parts of Europe. This changed in the 18th century—the introduction of new food crops and farming technologies increased the agricultural yield. There was also a providential decrease in epidemics during this time, though this was not due to any improvement in medicine which continued to be steeped in age-old therapeutics and theories.

Jethro Tull's seed drill

As human labor was expensive in the 18th century England, the English agriculturist Jethro Tull invented the seed drill in 1701, an important invention in the mechanization of agriculture. The seed drill—used for making holes in the ground, placing seeds in them and covering them up, three rows at a time— increased the rate of seed germination and thus the crop yield. Tull also invented a horse-drawn hoe, used for destroying the weeds and crumbling the soil, and improved the design of the plow. These early inventions were important steps in modernizing agriculture.

McCormick harvester

The shortage of labor during harvesting led to the introduction of harvesting technologies. Cyrus Hall McCormick, an American inventor, patented a successful horse-drawn reaper in 1834. The Great Exhibition in London in 1851 showcased the technology and soon the reaper was being used extensively, especially in the large farms of the American mid-west. This is a photograph of a worker driving a horse-drawn McCormick reaper while harvesting wheat.

Steam-powered traction engine

In 1769, Nicolas Cugnot introduced the first model of a traction engine. This design was modernized in the 1850s. These were steam-powered, wheeled engines which could be used to move heavy loads or for agricultural purposes. They were extremely heavy, slow and difficult to maneuver. Mainly used for providing power to move agricultural machinery by belt, they were also used for plowing where the conditions were right. The image shows a traction engine at work in a farm.

Agriculture saw many changes during the 17th and 18th centuries. New farm implements were introduced and a scientific study of agriculture and animal husbandry, which led to an increased understanding of the process of farming, was introduced. These changes brought about an increase in agricultural productivity which was important to feed a growing population.

In Europe and Britain, the Roterham plow, which had a Dutch design, was being used. In 1701, the English lawyer-turned-farmer Jethro Tull made many significant contributions to agricultural technology. He designed horse-drawn hoes which could destroy the weeds on the ground and also pulverize the soil so as to allow air and moisture to reach the roots. He also invented a seed drill which planted seeds in regular rows and thus reduced the need for fallowing. The rotary design used by him was subsequently used in many other implements.

The English inventor Robert Ransome devised a cast-iron share in 1785 and a self-sharpening share in 1803. The cast-iron plows were good for most soils except for the heavy black soil of the American prairies. In 1837, John Deere, a blacksmith in Illinois, made a plow from steel. This was a huge improvement since it allowed the cultivation of the heavy prairie soils and also required much less animal power. Another useful device invented in the 18th century was the mold plow which could drain wetlands.

Harvesting of crops being a seasonal activity, there was always a problem of finding enough local labor to bring in the produce. In 1784, a Scottish mechanical engineer, Andrew Meikle, invented the threshing machine to automatically remove the grain from the stalks. In 1831, Cyrus Hall McCormick, an American inventor, made the first automatic reaping machine whose design was later incorporated in many other machines. Around the same time, another American inventor, Hiram Moore, patented the first combined harvester. This was run by animal power and was very successful in areas where availability of labor during harvesting was a problem.

The next important innovation in agriculture was the introduction of steam power. After many attempts, the first usable steam plow was introduced in the 1860s and became very popular especially in areas where the land holding size was large. Several other agricultural implements using steam traction were also introduced around this time. More than in farming equipment the impact of steam was felt in transportation of agricultural goods. Distant markets could now be accessed easily for grain and livestock products. Cattle farming became a big enterprise in America. Railroads and steamships routinely carried agricultural goods to the markets. Tractors, which were originally run by steam power and subsequently by internal combustion engines, proved to be the workhorses of agriculture ever since their invention.

Fertilizers also played a big role in enhancing the agricultural productivity during this time. Although chemical fertilizers like saltpeter had been in use previously, others like sodium nitrate from Chile and guano from the Pacific coast of Peru were imported in large quantities. In 1842, the English agriculturalist John Lawes patented the synthetic process of producing superphosphate—by mixing powdered phosphate rock with sulfuric acid. This marked the beginning of the synthetic fertilizer era as nutrients were earlier made either from animal waste or mined minerals.

Most of the fertilizers which supplied nitrogen to the soil—an essential nutrient—were mined because it was difficult to manufacture ammonia, a precursor used in making the nitrogenous fertilizers. In 1908, German chemist Fritz Haber invented a process for synthesizing ammonia from nitrogen in the air and hydrogen. For the first time, nitrogenous fertilizers could be manufactured and this proved to be of great importance. As it turned out, ammonia could also be used in making munitions, a fact which was crucial during World War I.

Another achievement in agriculture during this period was the introduction of several new crops from the New World. Maize, tobacco, turkey, potato and cocoa were some of the major produce introduced from the Americas. Potato and corn were to prove to be of great importance. Imported to Ireland, potatoes became the staple food for the Irish and also soon became quite popular all over Europe. Similarly, corn was quickly adapted by farmers across Europe, especially in southern Europe.

Finally, another factor for the growth of agriculture was the beginning of scientific research and teaching in agricultural sciences. Agricultural colleges and departments within existing universities were established in England and Scotland, as well as in France and Germany from the middle of the 19th century. In 1862, the Department of Agriculture was set up in the US, which provided a further impetus to agriculture research and extension work.

Steam Power

Arguably the most critical invention of the 18th century was the harnessing of steam power. This technology was responsible for the Industrial Revolution and the mechanization of industries like textile manufacturing and mining, apart from revolutionizing transportation. The theoretical basis of the technology was the work of scientists in the 17th century who had conducted experiments to determine the nature of the vacuum, air pressure and heat. These included the English chemist Robert Boyle and the German scientist Otto von Guericke who had manufactured a vacuum pump in the 1650s.

However, the invention was left to the mechanical genius of the British engineer Thomas Newcomen who felt the need for replacing the expensive horses to remove water from coal mines. He experimented with many designs before finalizing the one which became the first commercially successful steam engine in 1712. The engine used the principle of condensation of steam to generate a partial vacuum

which then made the piston move. The piston was connected to a rod, which was attached to the pumping rod inside the mine. Although the engine was not very efficient, it was extremely useful in coal mines where the fuel was available at hand. Over the next few years, Newcomen's engine was used extensively in the coal mines in Britain.

Once the potential of a new form of power to supplement wind and water in industry and mining was realized, there were many improvements in the design of the steam engine over the next few decades. In 1765, a major innovation was made to the Newcomen engine by a Scottish instrument maker, James Watt. Watt dissociated the condensing chamber for steam so that the heating and cooling could be done separately. The condensing chamber was kept cold while the cylinder which housed the piston was kept hot. This proved to be a major improvement since it increased the fuel efficiency of the Newcomen engine. Watt also made another innovation: he designed an engine

Watt steam engine

Scottish engineer and inventor James Watt improved the steam engine to make it more efficient and capable of diverse applications. In 1769, James Watt patented a separate condenser for use with the Newcomen atmospheric engine, which had been invented in 1712 for use in removing water from mines. The condenser allowed much greater economy in terms of energy used to drive the engine. Over the next few decades, Watt further modified the design of the engine, and by 1800, Watt and his partner Matthew Boulton had sold more than 500 engines. This is a 1781 reproduction of the Watt steam engine. (France; 1781; Copper and glass; Conservotoire National des Arts et Metiers, Paris)

which could move a rotating shaft in a smooth rotary motion rather than the simple up-and-down motion possible with the Newcomen engine. While the earlier machine was only used for pumping water from mines, the rotatory engine could be applied to different machines—it was first used in 1783 in textile factories.

Spinning jenny

James Hargreaves, an English carpenter and spinner, made the first spinning jenny around 1764. Basically a multi-spooled spinning wheel, this device allowed a single spinner to weave many threads at the same time. The spinning jenny, together with rollers and moving trolley, improved the productivity of the textile industry dramatically.

Textiles

Watt's steam engine, which became hugely popular in the late 18th century, partly due to the monopoly he had on the patented design, was used in a cotton mill in what was the first application of steam power in manufacturing. It was also used for milling, replacing water power which had been used for centuries.

The transformation of the cotton manufacture from a domestic-based, small scale industry into a mechanized, factory-based one was a hallmark of the Industrial Revolution. The introduction of steam power was critical for this increase in the scale of manufacturing. But before that, there were many other innovations in cotton manufacturing which changed the way cloth was made and sold. The first step in mechanization of the loom was the introduction of the flying shuttle in 1733 by John Kay. It allowed the weaving of a much wider cotton cloth at a faster speed. Hitherto, the weaver's reach was the width of the cloth that could be woven on a loom. The shuttle increased

TECHNOLOGY TIMELINE

1701: The invention of the seed drill by the English agriculturalist Jethro Tull.

1712: Thomas Newcomen, a Baptist preacher and ironmonger, improves upon an existing design to make the first beam engine driven by steam.

1714: Mercury is used in a thermometer by the German scientist Fahrenheit. Fahrenheit was also the first person to devise a scale for measuring temperature.

1737: John Harrison, an English carpenter, invents the first marine chronometer.

1752: The American scientist and statesman Benjamin Franklin invents the lightning rod.

1764: Spinning Jenny, a multi-spool spinning device, is invented by James Hargreaves.

1765: The first practical steam engine is made by James Watt.

1783: The first hot-air balloon flight by the Montgolfier brothers in France.

1784: Edmund Cartwright builds the first power loom.

1793: The American inventor Eli Whitney invents the cotton gin, a simple device to efficiently separate cotton fibers from the seeds.

1796: Edward Jenner first demonstrates the use of vaccination against the deadly smallpox.

1801: The French silk weaver Joseph M. Jacquard invents the Jacquard loom which can weave very intricate

designs using punched cards. This idea was used later by Babbage in his analytical engine.

1804: The first steam locomotive built by the English inventor R. Trevithick.

1816: The English chemist Humphry Davy invents the Miners' Safety Lamp. The French physician R.T.H. Laennec invents the stethoscope.

1821: Michael Faraday builds the first electric motor using the force experienced by a current carrying conductor in a magnetic field.

1824: Joseph Aspidin, an English cement manufacturer, patents Portland cement.

1831: Faraday makes the first electric generator using the phenomenon of electromagnetic induction discovered by him. The American inventor Cyrus Hall McCormick invents the automatic reaping machine.

1834: Louis Braille, a blind Frenchman, perfects his Braille system for use by visually impaired people to read and write. Hiram Moore makes the first combined harvester.

1837: John Deere invents the steel plow. Samuel Morse patents the electromagnetic telegraph.

1839: Charles Goodyear invents the process of vulcanization of rubber to make natural rubber much harder and resistant to chemicals.

1842: The English agriculturalist J. Lawes patents a process to make superphosphate from sulfuric acid.

1856: The Englishman Henri Bessemer invents the Bessemer process to

make steel cheaply and in large quantities. Charles Parker invents a new material, cellulose, which would prove to be very useful in making photographic film later.

1860: Jean Joseph Lenoir makes an internal combustion engine running on lighting gas.

1862: French chemist Louis Pasteur invents the pasteurization process to prevent milk from fermenting.

1865: John Lister uses carbolic acid as an antiseptic on wounds.

1867: Dynamite is invented by the Swedish inventor Alfred Nobel. The typewriter is invented by Christopher Sholes, an American.

1876: Nikolaus Otto makes the first four-stroke internal combustion engine. Alexander Graham Bell patents the telephone.

1877: Thomas Alva Edison invents the phonograph to record and reproduce human voice.

1879: Edison makes the first long-lasting incandescent bulb.

1880: George Eastman patents a process of making dry plates for taking photographs.

1884: Gottlieb Daimler and Wilhelm Maybach develop a high speed, four-stroke engine burning gasoline. Lewis Waterman makes the first fountain pen.

1885: Daimler makes the first gas-engined motorcycle.

1888: John Boyd Dunlop develops the first pneumatic tyre. Heinrich

Hertz demonstrates the existence of radio waves.

1895: Auguste and Louis Lumière hold the first screening of a projected motion picture.

1895: Wilhelm Roentgen discovers X-rays and takes the first X-ray photograph of his wife's hand.

1901: The Italian inventor Guglielmo Marconi transmits a radio message across the Atlantic.

1901: Hubert Booth invents the vacuum cleaner.

1903: The Wright brothers fly the first heavier-than-air machine.

Edison's phonograph with boxes of cylindrical records

1904: John Fleming invents the thermionic diode valve. John Holt makes the first tractor for use in agriculture.

1906: Lee de Forest improves the diode valve to make a triode valve which can serve as an amplifier.

1910: Georges Claude makes the first neon lamp.

his width and, at the same time, speeded up the process significantly.

In 1738, Lewis Paul and John Wyatt invented the roller spinning machine and the flyer-and-bobbin system which used two rollers to draw cotton to even thickness; this led to the development of Richard Arkwright's water frame later. The hand-powered carding machine was introduced in 1748 by Lewis Paul for the wool industry. This was also the period when Watt's steam engine was being perfected and introduced into the textile industry.

The next great innovation came in 1764 when James Hargreaves, an English spinner and carpenter, invented the spinning jenny. This device allowed a single worker to produce much more thread. Arkwright's water frame, powered by water wheels, further increased productivity. In 1784, the power loom was introduced by Edmund Cartwright. This was the first use of mechanization, though the initial source of power was water. Later, steam power was used to run these power looms and a huge textile industry was established in Manchester.

The huge demand for cotton in the mechanized mills provided the impetus for the establishment of cotton plantations in the American South. In 1793, Eli Whitney, an American inventor, patented the cotton gin. This simple device made possible the efficient removal of seeds from the cotton fiber, a process which was very time- and labor-intensive previously. The gin was extremely simple in design and could be easily adapted to be used manually or with water or animal power. To a large extent, the gin was responsible for the tremendous increase in cotton cultivation in the American South, and hence provided the economic basis for slavery on the plantations.

Across the Atlantic, in 1804, the French inventor Joseph-Marie Jacquard, who had fought in the French Revolution alongside the revolutionaries, invented the Jacquard loom. This device allowed the weaving of very intricate designs such as tapestry and brocade. It made use of punched cards which allowed the weaver to weave any pattern automatically. The punched card technology played a vital role later on when it was adopted by the English mathematician Charles Babbage to design the first general purpose computer.

The invention of the sewing machine in the middle of the 19th century was another major milestone in textile manufacturing. This was originally made in France to produce army uniforms but the design was not very popular. In 1846, Elias Howe, an American, patented the lock-stitch design in which a curved needle carried the thread through the fabric, while on the other side of the cloth, it interlocked with another thread moving back and forth. This design was hugely successful, and in a few years, tens of thousands of sewing machines were being produced and used.

There were many other incremental process innovations in textile manufacturing in the coming years. Mercerization of cotton, introduced in 1850 by John Mercer, an English printer, was a very popular one. This involved applying sodium hydroxide on the cloth for a short time and then washing it away, a process which made the cotton take on brighter and longer-lasting colors. The zipper, a slide fastener which used spring clips, was introduced for the first time in 1893 and then improved upon subsequently. During the 20th century, it became popular as an attachment on many clothes.

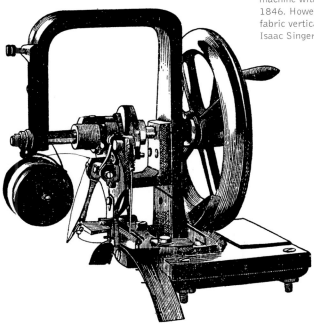

Transportation

It was not only textile manufacturing which underwent radical changes in this period, transportation also saw many innovations. The introduction of steam power in transportation changed the concept of distance. Railroads became the symbol of progress and the Industrial

Cotton gin

In 1794, the American inventor Eli Whitney was granted a patent for the cotton gin which automated the process of separating cotton fiber from seeds. A very simple device, the cotton gin used a combination of a wire screen and small wire hooks to remove cotton from the seedpods. This increased the productivity on cotton farms enormously—approximately 50 pounds of cotton could be produced in a day using the gin. As a result, Whitney's gin quickly gained tremendous popularity in the cotton plantations in the American South. This is a 1955 model of Whitney's gin.

Howe's sewing machine

In the 19th century, several people were working on the idea of a mechanical device for sewing. It was, however, the American inventor Elias Howe who patented the first sewing machine with a lock-stitch design in 1846. Howe's machine, which held the fabric vertically, was modernized by Isaac Singer in the 1850s.

Revolution. In the late 19th century, the introduction of the internal combustion engine proved to be as revolutionary as steam power was in the previous century.

The first use of the Watt engine in transportation was in steamboats. Robert Fulton, an American engineer, designed and ran a paddle steamer in 1807. Soon, steamboats became very common in America as well as Britain, but were used mainly for short haul routes since they required a huge amount of fuel. In the 1830s, for the first time, a steam-propelled boat was made to travel in the North Atlantic. All the initial boats were paddle boats and it was only much later that screw-propelled boats were introduced. Steamboats replaced sailing vessels all over the world by the end of the 19th century.

The Watt engine was improved upon by the English inventor Richard Trevithick who used high-pressure steam to drive the engines. Trevithick also built the first steam locomotive in 1804, though this was not very effective. The first successful steam locomotive was the famous Rocket developed by George Stephenson in 1829. This locomotive used horizontal cylinders beneath a boiler which was fired by coal. The Liverpool and Manchester Railway opened in 1830 and carried passenger and freight traffic on Stephenson's locomotives. The design of the Rocket was immensely successful and used for many years subsequently.

Initially, rails were made of cast iron; these were soon replaced by more durable wrought iron and then steel rails. The expansion of railroads occurred rapidly after this time, and in the next 50 years, railroads covered vast expanses of all continents. The penetration of railroads was of great significance in opening up huge marke for the manufactured products of the Industri Revolution as well as for providing raw materia for the hungry factories of the West. Th importance of the opening up of the America West with the expansion of the railways cann be overestimated.

If the steam engine was the workhorse of th 19th century, the internal combustion engin invented in the latter part of the 19th centur proved to be the mainstay of transportation in th 20th century. The steam engine, though a grea improvement over the existing sources of powe was not a very efficient engine. It was massive an had a very low efficiency—only a small part of th energy available in the fuel was converted to usef mechanical energy. The basic problem was that th fuel (coal) was burnt in a furnace which heate the water to generate steam. A need was felt eve in the early 19th century for a more efficient an portable source of power.

The scientific work which formed the basis c the design of heat engines was the theory of hea and thermodynamics formulated by scientist such as James Joule and Sadi Carnot. In fac Carnot, a French military engineer, wrote abou the construction of the ideal heat engine in 1824 This pioneering work was used by many engineer and inventors to try out various designs of engine which used the power produced by burning fue (including gunpowder) directly. However, thes attempts were not very successful.

In 1860, French engineer Etienne Lenoir mad an engine which worked with lighting gas. Th engine was a sort of modified steam engine; it wa not very efficient but was sold in large numbers fo

Stephenson's Rocket

George Stephenson, the principal inventor of the locomotive engine, was a railroad engineer who built many locomotives and several railroads. In 1829, along with his son Robert, he built the Rocket, a locomotive which won a competition of steam engines at Rainhill in October 1829. The Rocket could carry 30 passengers and reach a top speed of 36 mph (58 kmph). This is a replica of Stephenson's Rocket. (London; Science museum, South Kensington)

water pumps. In 1862, another French engineer Alphonse Beau de Rochas laid down the principles of what is really a four-stroke engine. This included a four-step process of intake, compression of the fuel mixture, burning of the mixture and expansion of the gases to power the piston, and finally an exhaust stroke to release the waste gases.

Many attempts were made to incorporate Rochas' ideas into a practical engine but were not very successful, till in 1876, Nikolaus Otto, a German engineer, built the first four-stroke engine. This engine was hugely successful and thousands were sold in the next few years. But the real advance came in 1884 when Gottlieb Daimler and Wilhelm Maybach, who had worked with Otto, developed a high-speed, four-stroke engine which burned gasoline. The essential requirement for this was the invention of the carburetor which used a combination of petrol and air to produce a flammable mixture that was injected into the cylinders to burn.

Karl Benz, a German engineer, made the first one-cylinder engine which ran on gasoline. Benz used it to run a three-wheeled automobile in 1885, while in the same year, Daimler and Maybach made the first motorcycle using Otto's design. They also powered a stagecoach and, in 1889, built the first

automobile with this engine; there was no looking back after this.

Around the same time that Daimler and Benz were using the Otto engine for making automobiles, another German thermal engineer, Rudolf Diesel, was trying to improve the efficiency of the Otto engine. The basic idea of Diesel was to eliminate the need for an ignition by an external agency and

Dunlop testing a pneumatic tire

John Boyd Dunlop developed a practical inflatable tire for his son's bicycle in 1888. The invention of pneumatic tires came at a crucial time since the internal combustion engine, which was to revolutionize road transport, was also being developed at this time.

TRANSPORTATION TIMELINE

1712: The first steam engine is made by Newcomen. This engine is a beam engine which can only have up and down motion. It is used extensively to remove water from mines. However, the concept of using steam as a source of power is introduced, which will have a tremendous impact on the future of transportation.

1737: John Harrison, an English carpenter, makes the first marine chronometer. This is an extremely useful device for navigators to determine their true longitude at sea.

1765: The first practical steam engine is made by James Watt, an instrument maker at the University of Glasgow. This engine could move a rotary shaft, thus allowing for its use for transportation.

1769: The Frenchman Nicolas Cugnot makes the first steam-powered tricycle, the world's first automobile.

1770: Modern bicycles invented.

1783: The first hot-air balloon flight by the Montgolfier brothers in France. The brothers, papermakers by profession, used a paper balloon filled with heated air.

1798: John McAdam invents a way to make roads with compacted stones. The roads built by this process, called macadam roads, were inexpensive and longer-lasting.

1804: The first steam locomotive built by the English inventor, R. Trevithick.

1807: Isaac de Rivas invents a hydrogen gas-powered vehicle.

1807: Robert Fulton, an American inventor, builds the steamboat using a modified Watt engine. The steam-powered paddle boat plied between New York and Albany.

1829: George Stephenson and his son make the Rocket, the first successful steam locomotive. The essential design of the Stephenson engine will be used in locomotives for many years to come.

1839: Charles Goodyear invents the process of vulcanization of rubber to make natural rubber much harder and resistant to chemicals.

1843: The first screw-propelled, large iron ship *Great Britain* is launched by I.K. Brunel. The ship served well for almost three decades. Brunel launched the *Great Eastern*, a huge double-hulled steamship, with both paddles and screw propeller in 1858.

1860: John Lenoir makes an internal combustion engine running on lighting gas.

1862: Alphonse Beau de Rochas propounds the principle of a four-stroke internal combustion engine.

1876: Nikolaus Otto invents the four-stroke internal combustion engine.

1869: George Westinghouse invents the compressed air locomotive brake, which evolved into the modern fail-safe air brake for trains.

1873: Andrew S. Hallidie invents the first cable car.

Wilhelm Maybach on a test run with Daimler's steel-wheeled automobile

1881: An electric-powered tricycle makes its debut in Paris.

1884: Gottlieb Daimler and Wilhelm Maybach develop a high speed, four-stroke engine which burned gasoline. They also make a carburetor which could be used to make the fuel and air mixture required for the internal combustion gasoline engine.

1885: Daimler makes the first motorcycle. Karl Benz makes the first practical automobile, a three-wheeled vehicle with a steel frame.

1886: Daimler makes the first four-wheeled vehicle to run on a high speed, four-stroke engine.

1888: John Boyd Dunlop develops the first pneumatic tyre.

1889: Daimler makes a rear-engine, four-speed automobile.

1900: The German cavalry officer Ferdinand von Zeppelin makes the first flight of an airship. The rigid airships or zeppelins were immensely successful and were used till the late 1930s despite the airplanes. The most famous one, the *Hindenberg*, exploded in 1937.

1903: The Wright brothers invent and fly the first heavier-than-air flying machine, a 40 ft (12 m) biplane powered by a 12 hp engine.

1906: The American brothers, Francis and Freelan Stanley, make a steam powered automobile which makes a world record of traveling a mile in 28.2 seconds, a speed of over 120 mph (193 kmph).

1907: The Breguet brothers make a flight using a gyroplane, a flying machine which used four rotors. The gyroplane was the precursor of the helicopter.

1908: The American entrepreneur Henry Ford builds the first automobile for the mass market. Model T is immensely successful as it is inexpensive, versatile and easy to maintain.

1908: Hydrofoil boats invented by Graham Bell and Casey Baldwin.

instead use the compression of the piston to ignite the fuel. Diesel tried using coal dust as fuel but this was not very successful. Finally, liquid petroleum was used and soon the first commercial diesel engines were being built and sold, and were used extensively in submarines and ships.

Another invention which was of importance in the development of automobiles was that of the pneumatic tire. In 1845, leather tires with air were invented by Robert Thompson in England but these never became very popular. It was only in 1888 that a veterinary surgeon, John Boyd Dunlop, developed pneumatic or air-filled tires for bicycles. The pneumatic tire, in which the load is taken by compressed air, proved to be very useful in reducing friction in vehicles and thus quickly became very popular.

The automobiles and the railways changed the concept of distance for human beings in a fundamental way. Now for the first time, expanses could be bridged for men and material. Besides the developments in land and sea transport, this period also saw humans taking to the skies for the first time.

The Wright Flyer II biplane

In 1903, the Wright brothers, Orville and Wilbur, carried out the first successful flight of a powered, controlled and heavier-than-air machine. The brothers achieved this remarkable feat after years of experimentation with gliders and control mechanisms. They continued to improve their designs and, finally in 1905, unveiled their most successful flyer which stayed aloft for over 30 minutes.

The human fascination with flying has been as old as civilization. The myth of Icarus is just one example of humans imitating birds to take to the skies. In 1783, the Montgolfier brothers in France showed that a balloon filled with hot air would rise up in the air. Soon, after sending various animals up in the hot-air balloon, the first manned flight took place in France. Over the next few years, many modifications were made to ballooning, the most notable being the use of hydrogen to get a much more powerful lift. Ferdinand von Zeppelin, a German soldier, added a rigid, though light, frame to the balloon to make it steerable. These dirigibles, or zeppelins, were used for carrying passengers and played a major role during World War I.

Apart from balloons and zeppelins, gliders were also being investigated in the 19th century. In 1891, Otto Lilienthal in Germany built the first glider capable of carrying a human being. His design was improved upon by the American engineer Octave Chanute, who fabricated very stable and sturdy gliders.

Among others experimenting with gliders were two brothers in America, Orville and Wilbur Wright. The brothers, who were bicycle makers, made their first successful glider in 1902 and continued their work on building flying machines. In 1903, Orville Wright became the first human being to fly in a heavier-than-air machine. The 40-ft (12-m) biplane was powered by an engine and flew for over 100 ft (30 m) in about 12 seconds. In the coming years, the Wright brothers perfected a steering mechanism for their machines and, finally in 1905, obtained patents for their inventions.

Metallurgy, Mining and New Materials

Though the Industrial Revolution was primarily based on steam power and automation, innovations in metallurgy, mining and new materials proved to be of vital importance. These included an economical and efficient method of manufacturing steel and cement.

By the end of the 17th century, there was a severe depletion of forest resources in Britain since wood and charcoal were used for smelting iron from ore as well as for making steel. In 1709, an ironsmith, Abraham Darby, was successful in using coke to reduce iron from its ores. This was an important step since it allowed the use of the large coal resources of Britain to fuel the huge demand for iron and steel. In 1740, the process for making crucible steel, which led to the production of high quality steel, was developed by Benjamin Huntsman. In the 1780s, a new process to make wrought iron from cast iron efficiently was discovered.

The easy availability of cast iron—used for making bridges, aqueducts and other construction—and wrought iron—used for making instruments and machinery—played an important role in the expansion of the Industrial Revolution in Britain in the 18th century. But the real revolution came with the invention of a cheap and fast method to make steel in large quantities in the middle of the 19th century.

In 1856, Henry Bessemer, an Englishman, was attempting to make a stronger cast iron to manufacture a canon which was robust enough to use an improved artillery shell invented by him. One of the techniques he discovered was that the oxygen in the gases emanating from his furnace could remove carbon from the pig iron. He then pumped air through the molten cast iron and this purified the iron and also heated it to reduce the viscosity so that it could be poured easily.

Bessemer perfected these techniques in what has come to be known as the Bessemer process and was able to produce large amounts of pure wrought iron. He also made a converter—a huge oval container made from steel and lined with clay or limestone which is pivoted on huge stands. The iron is put in from the top and air is pumped in at high pressure from holes in the bottom. The Bessemer converter used very little coke and could make a large quantity of steel in a very short time. Later, there were innovations to remove other impurities like sulfur and phosphorous, and soon high-quality steel was obtained which could replace wrought iron in most applications. The availability of high quality, inexpensive steel proved to be a boon for the economy.

With the availability of cheap steel and the demand for machinery to move the Industrial Revolution, there was an expansion of the mechanical engineering industry to manufacture

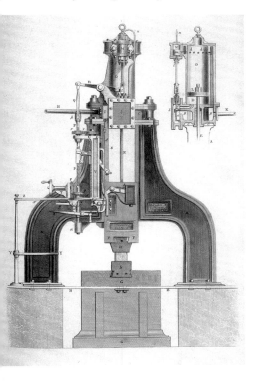

heavy and precision machinery. The lathe was now steam-powered and much more accurate and efficient than the previous versions. James Nasmyth invented the steam hammer which, together with better drilling machines, became an important addition to the workshops.

The invention of the safety lamp by the English chemist Humphry Davy in 1815 proved to be of great importance since it increased the safety of the miners working in the deep mines. The presence of inflammable gases like methane in these mines had previously made mining very dangerous and uneconomical. This simple modification of the traditional miners' lamp made a great improvement in the quality of miners' safety.

This period, especially after the invention of steam power, also saw many developments in the field of civil engineering. Tools which used hydraulic power and compressed air increased the efficiency to a large extent. Cast iron was replaced by wrought iron and later by steel in bridge-building. The invention of Portland cement in 1824 by

Joseph Aspidin was a major step in construction. This material, when mixed with water, produced concrete which, when used with steel, could provide unprecedented strength to the buildings.

The major breakthrough though in civil engineering was the invention of dynamite in 1867. Alfred Nobel, a Swedish physicist, was experimenting with nitroglycerin—a dangerously explosive chemical—when accidentally he hit upon the idea of mixing it with kieselguhr—a kind of porous earth. The resulting mass was much safer than nitroglycerin and could be used for engineering as well as military purposes.

Another material which was discovered in the 18th century was rubber. Natural rubber was first discovered by the French in the jungles of South America. Although it was useful because of its unique properties, its use was restricted till Charles Goodyear of America invented the process of vulcanization of rubber in 1839. Vulcanized rubber, which was obtained by adding sulfur to rubber at high heat, was harder and more resistant to chemicals, and could now be used in a variety of goods and machinery.

Alfred Nobel

Swedish inventor, chemist and philanthropist Alfred Nobel invented the blasting cap which initiated the use of explosives for purposes other than warfare. He also invented dynamite by mixing the extremely volatile nitroglycerine with an inert substance kieselguhr. Nobel made a fortune from this invention, which was much safer and easier to handle than any other explosive.

Steam hammer

In 1839, the English engineer James Nasmyth invented the steam hammer to forge heavy metal pieces. Devised to forge the drive shafts for the paddle wheels of the steamship *Great Britain*, the steam hammer was later used in all large workshops. Nasmyth also manufactured more than a hundred locomotives, a steam ram, a hydraulic press and many other machines. (*Cyclopedia of Useful Arts* vol II)

Communications

The field of communications saw revolutionary developments in the 19th century. The invention of the movable type and the printing press by Gutenberg had already spurred the growth of printing. Developments in printing technology and papermaking during the 18th and 19th centuries were crucial to further enhance the reach of the printed word. The high-speed rotary press was an important innovation which speeded printing enormously. In 1884, Ottmar Mergenthaler in the US patented the linotype—a typesetting machine which allowed casting of complete lines rather than individual letters, thus speeding up the composing process significantly.

Typewriters were invented in 1867 by the American inventor Christopher Sholes who improved upon his initial design in 1868 and got a patent in the same year. In 1873, he started collaborating with E. Remington and Sons for the commercialization of typewriters, and soon they were being used all over the US.

In all this mechanization, writing technology also underwent some changes. The fountain pen was invented in 1884 to replace the instruments which had to be dipped in the inkwell. Invented by Lewis Waterman, the fountain pen used the capillary action for the ink to flow to the nib. This was followed by the development of the ballpoint pen in 1895, though its design was not very satisfactory. A successful design was introduced by László Biró, a Hungarian, in 1931.

But the biggest revolution was in imaging. The invention of photography was in some senses as important as the introduction of movable type a few centuries ago. In 1820s, Joseph Nicéphore Niépce took the first photograph using a pewter plate with bitumen. Over the next few years, his partner Louis Daguerre improved upon the photographic process by using light-sensitive silver chemicals,

and by the 1850s, daguerreotypes were bein used extensively.

In 1880, the American inventor George Eastma patented a process of making dry plates fo taking pictures. He also introduced easy to us Kodak cameras in 1888—these were handhe cameras with a roll of film. This was a major ste in popularizing photography. The film was base on celluloid, a material invented in 1856 by th Englishman Alexander Parkes but not popularize till the 1870s.

The telegraph was another invention whic changed communications in a big way. Optic telegraphy, most notably the semaphore syste

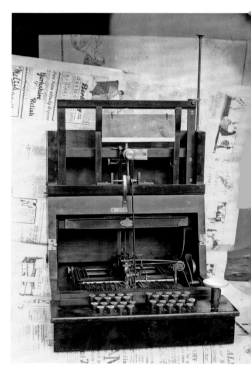

invented in France in the 18th century, was alread in use when electrical telegraphy was invente The work done in the production of electricity an

Early typewriter

After many unsuccessful attempts in the 19th century to make a practical typewriter, the American inventor Christopher Sholes invented a writing machine whose speed exceeded that of a pen. It was a crude machine which went through several improvements in subsequent years before the first typewriters were sold in 1874. This 1866 model was a forerunner of the typewriter patented by Sholes in 1868.

Morse code

The American painter Samuel Morse developed the first electric telegraph and the Morse code. A system of dots and dashes was used extensively throughout the world for telegraphy. The words of the first message sent by Morse code—from Baltimore to Washington in 1844—were: "What hath God wrought!" Seen here, a telegraph receiver printing Morse code onto ticker tape and a slate displaying the code.

also studying its magnetic properties set the stage for the invention of the first practical telegraphy system by the British scientists William Fothergill Cooke and Charles Wheatstone in 1837. While they were developing the telegraph in England, an American professor of painting had developed a very useful system of representing letters by dots and dashes. Samuel Morse got a patent for the electromagnetic telegraph in 1837. Over the next few years, he improved on his original design till, in 1844, a message was transmitted and received over a distance of 37 miles (60 km) between Washington DC and Baltimore. The telegraph era had begun—over the next few years this became the dominant mode of speedy communication used extensively in the railways, newspaper industry and stock exchanges.

If the telegraph was the most important technological innovation in communicating the written word, the invention of the telephone in 1876 was to revolutionize oral communication. Alexander Graham Bell, a Scottish-American inventor, was the first person to be granted a patent for transmitting speech sounds over electrical wires in 1876. Bell's work was built on the well understood principles of sound and electricity and his patent is considered to be the most valuable one ever granted.

After Bell's invention of the telephone, several inventors like Thomas Watson and Thomas Edison worked on the design of the telephone and, by the end of the century, the design used for the next five decades emerged. In 1878, the first telephone exchange was established in Connecticut which connected 21 subscribers. This was a manual exchange—it was only in 1889 that an automatic switching mechanism was introduced.

In 1877, the prolific American inventor Thomas Alva Edison produced the first working phonograph. This was a device to record and replay sound using electricity. The first phonograph was made by making indentations on a tinfoil wrapped around a cylinder. The indentations were made by a stylus which moved in response to sound waves. These indentations then could be replayed with another stylus which converts them back into sound.

Telegraph and telephones were undoubtedly revolutionary technologies and had a huge impact on history. But if the 19th century was the age of telephone and telegraph, the next century saw the development of another technology which overtook these in its usefulness. Radio communication was a direct result of the theoretical work carried out in the middle and late 19th century in the field of electricity and magnetism. The English scientist James Clerk Maxwell had worked out a comprehensive, unified theory of electricity and magnetism in 1864. One of the predictions of

his theory was the existence of electromagnetic waves which could travel in a vacuum with the speed of light. In fact, light was just one kind of electromagnetic wave.

In 1888, the German scientist Heinrich Hertz demonstrated the existence of these waves over a very short distance in the laboratory. But it was the young Italian inventor Guglielmo Marconi who improved upon Hertz's results and showed how signals could be transmitted over much larger distances. The crowning achievement was in 1901 when Marconi was able to transmit a wireless message across the Atlantic. The beginning of the new century heralded the era of wireless communication.

Progress in the field of wireless communication was dependent on the development of electronic circuits and devices. The most important invention which enabled the growth of radio communication was that of the diode valve in 1904 by an English electrical engineer, John Fleming. Fleming's device could act as a valve in the sense that it allowed the flow of current only in one direction. This was followed up by the invention of the triode by Lee de Forest in 1906 which revolutionized radio communication. The triode could be used as

an amplifier and this allowed communication even with weak signals.

The radio was initially used primarily for wireless telegraphy and its potential was demonstrated dramatically in the SOS signal sent out by the sinking RMS *Titanic* in 1912. In 1915, AT&T demonstrated for the first time that speech could be transmitted using wireless; there was no looking back for radio technology after this.

Electricity

Although steam power was the driving force behind the Industrial Revolution in the late 18th century, electricity proved to be the technology with a much larger and long-term impact.

Marconi transmitting wireless signals

The Italian engineer and physicist Guglielmo Marconi improved on Hertz's apparatus to generate and receive radio waves. By 1901, Marconi was able to transmit and receive messages in Morse code across the Atlantic. He continued to improve the apparatus, which led to the invention of the first radio broadcast in 1906 by Reginald Fessenden. Here Marconi is seen working on an apparatus similar to the one with which the first signals were sent across the Atlantic.

Franklin's lightning experiment

American writer, inventor, scientist, diplomat, printer and publisher Benjamin Franklin made many contributions to science, especially in the field of electricity. He was the first person to demonstrate that lightning was really electricity by a famous experiment of flying a kite in a rainstorm, depicted in this lithograph.

Electric inductor

Michael Faraday, inventor of the electric dynamo and the motor, contributed to the fields of electromagnetism and electrochemistry. Faraday used a crude electric inductor to discover electromagnetic induction and the law governing it. This enabled the invention of the electric generator and transformer.

Electrical phenomenon had been studied as early as the 17th century when the English scientist William Gilbert demonstrated the attractive force experienced by objects when they are rubbed. Subsequently in 1750, the American inventor and politician Benjamin Franklin carried out his famous experiments with lightning to demonstrate that it was electricity.

Several scientists worked to understand the nature of electricity and its properties in the latter part of the 18th century. The Italian physician Luigi Galvani demonstrated the presence of electricity in the transmission of nerve signals in 1766, which was followed by the work of the Italian scientist Alessandro Volta, who made the first battery in 1800, called the voltaic pile, by placing two dissimilar metal rods in brine.

But it was a bookbinder's apprentice, Michael Faraday, who discovered the most important properties of electricity. Faraday carried out studies on the nature of electricity and in the 1820s made the first electrical motor—a device which for the first time transformed electrical energy into mechanical energy, using the interaction of electric current and magnetism. Faraday continued his work on electricity and discovered the phenomenon of electromagnetic induction, or the production of electric current from the change of a magnetic field. This led him to make the first dynamo in 1831, a device which converted mechanical energy into electrical energy.

The design of the electrical motor and the dynamo both underwent improvements in the next few years, resulting in larger motors capable of replacing steam power as well as huge generators to generate electric power. Electricity transformed

life in ways in which few technologies had done in history. Economical electrical lighting, electric tramways and the use of electrical power in manufacturing were all made possible before the end of the 19th century.

The use of electricity for lighting began in the early 19th century. The first lamps were the electrical arc lamps—these used carbon electrodes between which an electrical arc was struck to generate light. These lamps, though difficult to

operate, proved to be very popular, especially in street lighting. By the middle of the century, several scientists were experimenting with the use of filaments for lighting. An Englishman, Joseph Swan, used carbon filaments and even cotton thread dipped in acid to pass an electric current and generate light.

It was left to the genius of the American inventor Edison to put all the ideas together to patent the first incandescent lamp. He used a vacuum pump to create a near vacuum in a glass bulb fitted with carbon wire. The vacuum was needed to ensure a longer life for the filament. Swan too was trying out various materials for use as filaments and there is a controversy regarding who actually should be credited with the invention of the incandescent bulb. In 1879, Edison made the first long-lasting lamp with carbonized thread which glowed continuously for over two days. After this demonstration, many other materials were tried out for longer-lasting filaments. The first commercial use of the incandescent lamp was on the ship *Columbia* in 1880 and, over the next few years, incandescent bulbs were being used in factories, merchant establishments and

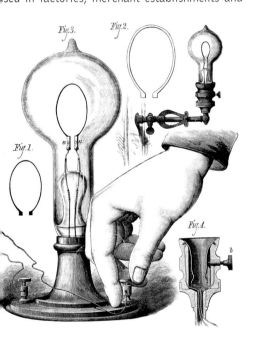

homes. If the railway had changed the concept of distance for humans, reliable, artificial lighting in the form of incandescent bulbs liberated humans from the cycle of day and night as far as work was concerned.

In 1907, the first tungsten filament lamps—which were brighter and longer-lasting than the carbon filament lamps—were introduced by Franjo Hannaman. Till very recently, most of the incandescent bulbs used tungsten filament. Several design modifications were made in the next few years, including introducing rare gases in the bulb to increase the life and using a coiled filament.

The other lighting technology that was invented around the turn of the century was that of electric discharge lamps. In 1901, a mercury vapor lamp was introduced and it gained rapid popularity.

The basic principle in an electric discharge lamp is to use two electrodes separated by a gas to pass an electric current. In 1910, the first neon lamp was made by Georges Claude, who applied a high voltage to a tube with neon gas. The light was red in color and the commercial potential of this development was soon realized. Neon lamps were soon being used in signboards and billboards around the world.

Medicine

The science and practice of medicine saw many developments during the 18th and 19th centuries. Though there had been a lot of progress in medicine in the Middle East during the Golden Age of Islam in the 10th and 11th century AD, the practice of medicine in the West had essentially been empirical with little theoretical basis.

In 1796, Edward Jenner, an English physician, used the discharge from cowpox sores to inoculate a young child against smallpox. It was a remarkably successful experiment—even after he exposed the patient to smallpox, there were no symptoms. Vaccination against infectious diseases thus started and proved to be one of the most beneficial inventions in history. Millions of lives have been saved since then because of extensive use of inoculation.

In 1819, the French physician R.T.H. Laennec, used a hollow wooden cylinder to listen to the heart sounds from a patient's chest. This was the first use of what came to be known as a stethoscope. Further refined with rubber tubes and two earpieces, the stethoscope became an essential tool for physicians and continues to be used till today.

The big breakthrough in the theory and practice of medicine came with the work of the French chemist Louis Pasteur in the middle of the 19th century. Pasteur was researching the process of fermentation which was of great importance to the wine industry in France. In a set of very elegant experiments, Pasteur showed that fermentation in milk and wine was a result of presence of certain microscopic organisms. If one ensured that these micro-organisms were absent, then fermentation will not take place. There was still the question of whether the micro-organisms were being spontaneously generated or whether they developed because of contamination. Pasteur proved that it was due to exposure to air that germs got into the substance. This revelation led to methods of controlling the spread of germs. He also developed the process of heating milk to prevent its fermentation, a process which is now called pasteurization.

Building on the work of Edward Jenner, Pasteur successfully used inoculations to prevent a deadly disease, anthrax, in sheep and cattle. His major

Edison's incandescent bulb

When Thomas Edison started his experiments with the electric bulb, inventors had been working on producing a practical, inexpensive and long-lasting electric light for over 50 years. Working with his assistant Francis Upton, Edison tried out many designs till he hit upon the idea of using carbonized bamboo fiber as the filament in 1880. The electric bulb, over the next few decades, replaced gas lighting prevalent in homes and commercial establishments.

contribution in this field was the development of a vaccine against the fatal disease rabies, which was caused by animal bites.

Pasteur's work on the spread of disease-causing germs led to a wider acceptance of the germ theory of disease. In 1865, the Scottish surgeon Joseph Lister used carbolic acid on open wounds to control infections. This was the beginning of antiseptics. In 1882, a German doctor, Robert Koch, managed to isolate and study the bacteria causing tuberculosis as well as cholera. In the next few years, many other diseases were linked to the presence of micro-organisms.

The use of anesthesia in surgery was another major innovation in the middle of the 19th century. In 1846, the use of ether as an anesthetic was demonstrated for the first time in the US and, in a few years, it was being used in many places. In 1847, chloroform was introduced; it was much safer than ether and became the anesthetic of choice.

Toward the end of the century, a British military doctor, Ronald Ross, working in India discovered the cause of malaria and how it was spread—he found that the anopheles mosquito carried the parasite in its stomach. Ross' work was crucial in controlling malaria by protecting against mosquito

bones of Roentgen's wife's hand. X-rays proved to be of great use in diagnosis and treatment especially in orthopedics, and turned out to be an invaluable tool for studying the composition and structure of materials.

The 19th century was thus a time of great significance in the history of medicine since many breakthroughs were achieved in this period. The germ theory, which was now well established, provided the theoretical foundations of public health practices, which led to a huge decrease in mortality and morbidity. The enhanced knowledge about the causes of diseases, the development of vaccines for infectious diseases as well as a better understanding of human physiology set the stage for the revolutionary developments that were to come in the next century.

Enabling Inventions

Another distinctive characteristic of the period from the 18th to the 20th century was the development of many "enabling" inventions and discoveries. Some of the inventions of this period were not great achievements in themselves but these enabled the invention and discovery of other things and concepts. For instance, the German thermometer maker Daniel Gabriel Fahrenheit first devised a scale for measuring temperature in 1724. He was also was the first person to use mercury in a thermometer. The temperature scale and the thermometer were very important discoveries and played an important part in the subsequent development of the science of thermodynamics, which in turn formed the basis of steam power as well as the internal combustion engine.

Similarly, the development of the theory of electricity and magnetism played an important part in the 19th century. Electric power (used in illumination, transport and manufacturing and so on), telegraph, telephone and radio—these were all based on the theoretical understanding of electrical and magnetic phenomenon.

After the theory of electromagnetism was propounded, what remained to be understood was the nature of matter at the microscopic and submicroscopic level. The English schoolmaster John Dalton proposed a consistent theory of the atomic structure of matter in the early years of the 19th century. Dalton's atomic theory was much more refined than the early ideas about indivisible particles making up matter, first referred to by the ancient Greeks and the ancient Indian philosophers. Dalton consolidated his studies in *New System of Chemical Philosophy*.

The understanding of electricity in the middle of the 19th century, primarily due to the work of Faraday and James Clerk Maxwell, had led many

X-ray or Roentgen tube

In 1896, the German physicist Wilhelm Conrad Roentgen discovered an unknown or "x" radiation that could pass through wood, cloth, paper, or even skin. A week later, he took the first X-ray photograph of his wife's hand, showing the bones of her hand. As the medical implications of this form of body imaging were soon realized, X-ray was hailed as one of the most important inventions in medicine.

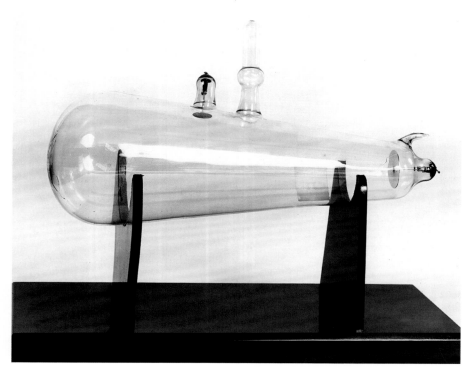

bites. Malaria was the major cause of morbidity in the swampy areas; this could now be prevented by adopting protection against mosquitoes.

In diagnostics, the major breakthrough came with the work of the German physicist Wilhelm Roentgen, who in 1895 accidentally discovered X-rays and saw the potential of X-rays to image bones. The first X-ray photograph showed the

Marie Curie in her laboratory

The Polish chemist Marie Sklodowska Curie was the first woman to win a Nobel Prize in 1903. Her contribution to the field of radioactivity and discovery of radium would help arm humankind against diseases such as cancer. She was the first person to receive the Nobel Prize twice, and the only one till today to win it for two different sciences. During World War I, Marie Curie set up X-ray vans, training 150 female manipulators to ensure these mobile radiography units, called "Little Curies", reached the wounded in time. She also donated the gold from her Nobel Prizes toward the war effort.

cientists to conjecture about the constitution of lectricity in terms of discrete units. In 1897, .J. Thompson, a scientist at the Cambridge University, discovered the first sub-atomic article: the electron. A whole new world, of he submicroscopic particles, opened up for nvestigation with this discovery.

In 1896, Henri Becquerel discovered that ertain minerals gave out a mysterious radiation which caused photographic plates to darken. This iscovery of the phenomenon of radioactivity was mportant toward understanding the structure of he atom. Subsequent work by the Polish chemist Marie Curie and her French husband Pierre Curie, s well as by the New Zealander Ernest Rutherford, larified the nature of radioactivity and also played vital role in unraveling the structure of matter at ne atomic level. The work of the Curies provided n effective cure for diseases like cancer.

The first decade of the 20th century saw the birth f one of the pillars of modern physics, namely Albert Einstein's Theory of Relativity. His equation $E=mc^2$ led to the development of the atom bomb, nleashing a new form of terror in the universe. In 905, Einstein, working in the patent office, wrote hree seminal papers in physics. In one of them, he roposed a radically new theory about the nature f space and time. In another one, he explained the ewly discovered photoelectric effect by proposing new approach to the nature of light. In this, he was building on the ideas of Max Planck, who in 900 had proposed that energy (or light) took the orm of packets called quanta.

Relativity and later quantum mechanics became he two foundations of modern physics. This mproved understanding of the nature of matter,

energy and space led to many technological developments in the 20th century.

The period from 1700 to 1914 was thus one of the most progressive times for humanity in terms of technological and scientific advancement. The Industrial Revolution, the speeding up of transport and communications, the developments in medicine and agriculture, all led to a dramatic change in the lives of a large percentage of the world's population.

But the pace of change experienced in the two centuries was nothing compared to what the world would see in the coming century. The innovations in the 20th century would be almost as radical as the transition from hunting-gathering to settled agriculture or the introduction of steam power.

Albert Einstein

Possibly the most well-known scientist in the 20th century, Albert Einstein created a sensation by demolishing Newtonian principles by his Theory of Relativity. When Einstein published the equation $E=mc^2$, he did not foresee that it would lead to the invention of the atomic bomb. Fearing that the Germans would make the atomic bomb, Einstein, in a letter to President Roosevelt in August 1939, described the potency of atomic energy and urged him to build the bomb before the enemy. Much to his regret, the bomb was used by the Americans against Japan.

Inventors of the Period

Most scientific and technological development is seldom single-handed work of an individual since it usually builds on the work of several others. However, there are some individuals who are trailblazers in the sense that their contributions are seminal and open up new vistas in their respective fields. In this light, the 18th and 19th centuries saw many remarkably innovative individuals who were responsible for making this period exceptional in terms of technological advancement.

The critical invention of the 18th century was the steam engine which revolutionized mining, transportation and manufacturing. James Watt, an instrument maker in Glasgow, was responsible for making the first practical steam engine. It was a huge improvement on the existing Newcomen engine which had been invented a few decades earlier. Watt's engine had a separate chamber for condensation and could also move a rotating shaft. Watt did not have the capital that was required to produce the engines or get a patent on his design. He formed a collaboration with a foundry owner, Matthew Boulton, to manufacture and sell the engines. The steam engines sold very well and soon found extensive use in mining and later in textile factories.

Michael Faraday

One of the greatest scientists of the 19th century, Faraday started his career as an apprentice to a bookbinder. He got interested in problems in chemistry and later in electricity. Faraday made fundamental contributions to both electrochemistry and electricity. He is possibly the only scientist to have his picture on a banknote.

If Watt's invention enabled the Industrial Revolution, then Michael Faraday's work on electricity and magnetism was responsible for the introduction of another source of power: electricity. Faraday was the son of a poor blacksmith who had to leave school to be a bookbinder's apprentice. This gave him an opportunity to read many books including those on chemistry. He started working as a secretary to the famous chemist Humphry Davy and did extensive work in many areas of chemistry. However, it was his work in electricity—which he started after hearing of the Danish scientist Oersted's experiment on the interaction of electricity and magnetism—that distinguished him. Faraday invented the first electric motor in 1821 (though there was some controversy about others having done this before him) and this was the first demonstration that electricity could be used to perform work.

Later in his life, Faraday discovered electromagnetic induction, possibly his greatest achievement since this provided a way to generate electric power in a reliable and lasting way. Hitherto, electric current could only be produced using voltaic piles—a primitive form of batteries—but with Faraday's work it was possible to produce electricity using any form of mechanical power. Faraday constructed the first dynamo to practically demonstrate that this could be done. Faraday's work on electromagnetic induction was also an important step for another Englishman, James Maxwell, to formulate a unified theory of electromagnetism, which ultimately led to the development of the radio.

Though the inventions of men like Faraday and Watt were revolutionary in nature, it was really the "Wizard of Menlo Park", Thomas Alva Edison, who has come to epitomize the innovative spirit in the late 19th and early 20th century. Born in a humble family in Ohio, in childhood he sold newspapers on trains and became a telegraph operator as a teenager. His inventive streak was evident even at an early age as he invented several devices related to the telegraph. He improved the stock ticker in 1869 for which the Gold and Stock Telegraph Company paid him 40,000 dollars.

He moved to New Jersey where he was to spend the rest of his life working on many inventions. His first big success was the invention of the phonograph in 1877. Though crude in many ways, it proved to be a big hit because of its novelty. Edison set up an industrial research laboratory, the first of its kind, in Menlo Park where he employed several engineers to work on his ideas.

Edison and his engineers invented several devices including the carbon microphone which became standard in the mouthpieces of telephones. But his next big invention was a practical incandescent light bulb. Light bulbs had been invented earlier but the design was not long-lasting. Edison's engineers worked on several materials for the filament of the bulb till in 1879 he discovered that carbon filament gave a longer-lasting bulb. He formed the Edison Electric Light Company and promoted the use of electric lights in cities and homes. At the same time he improved and invented various systems of electric generation and distribution.

Over his lifetime, Edison obtained an astonishing number of patents—more than a thousand. But that was not the only reason he is considered to be the inventive genius—it was also because the nature of his inventions was such that they made instant connection with the public. The phonographs, the bulb, the motion picture camera, the Kinetoscope were all inventions which gave him immense popularity and fame.

Another American inventor who became enormously famous during his lifetime was Alexander Graham Bell. Born in Edinburgh, Bell moved to the US at the age of 24 and started working with the deaf. In his spare time, he experimented with improving the designs of the telegraph.

The telegraph, by the late 19th century, had become very widespread. In fact, it was really the nervous system of commerce and business. Its impact can be compared to the impact of the Internet today. However, the telegraph system was still in need of major innovations, especially in finding a way to send simultaneous multiple messages on the same wire rather than installing new wires. Several inventors, including Edison, were working on solving this problem.

Bell started working on a method of transmitting human voice via the telegraph together with his assistant Thomas Watson. In 1876, Bell transmitted the now famous words "Mr. Watson, come here, I want to see you" to his assistant in an adjoining room. The telephone became a reality with these few words. Soon, Bell demonstrated that his device could be used over longer distances and offered to sell his patent to the Western Union Telegraph Company, which refused the offer saying the device was only a toy! As it turned out, Bell and Watson retained the patent and the Bell Telephone Company that they founded became one of the most successful in the world on the basis of his patent.

The Italian-Irish inventor Guglielmo Marconi was another person whose inventions brought about fundamental changes in the way we communicate. Born in Italy, Marconi developed an early interest in electricity while still a teenager. He began experimenting with Hertzian waves, as radio waves were then known, and in 1897, while still a young man of 23, demonstrated the transmission of telegraphic messages wirelessly over a distance of 4 miles (6 km). Subsequently, in 1902, he managed to send a wireless transmission across the Atlantic and proved that it was feasible to replace expensive wired telegraph with his system of radio communication. Though the use of radio waves to devise a system of wireless telegraphy had been in existence before Marconi, his achievement lay in trasmitting these waves over long distances, paving the way for the invention of the radio. Marconi got the 1909 Nobel Prize in Physics, with Karl Ferdinand Braun, for their pioneering work on wireless telegraphy.

The period from 1700 to 1914 can be called the age of inventors since many individuals brought about major inventions during this period. Some of them broke new ground with their ideas while others improved upon existing ideas and inventions to make them practical and commercial. Most inventors, especially in the US, were entrepreneurs who made enormous fortunes with their inventions. These included Bell, Edison, the Wright Brothers, Bessemer and James Watt among others. However, this period of individuals working alone in their laboratories or workshops and inventing was soon to end.

The 20th century will see the rise of the industrial laboratory as well as huge government sponsored laboratory infrastructure in institutes and universities where most of the research and development would take place. Edison's foresight in setting up the industrial laboratory in Menlo Park was remarkable in this regard. In a way, it predicted the future of innovations.

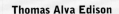

Thomas Alva Edison

Thomas Alva Edison was an American inventor who held the record for a total of 1093 patents. Edison is credited with inventing several devices, including the first practical bulb, the phonograph, the first motion-picture apparatus and the microphone. He also set up the world's first industrial laboratory: the Invention Factory in New Jersey. Here Edison is seen with one of his most important inventions, the phonograph.

Alexander Graham Bell

The Scottish audiologist and inventor Alexander Graham Bell made many contributions to teaching speech to the deaf before inventing the telephone in 1876. After this innovation, Bell continued his experiments and invented the photophone, a precursor to modern fiber optics, in 1880. Here Bell is seen inaugurating the New York–Chicago telephone line.

THE WAR PERIOD
DEVELOPMENT AND DEVASTATION (1914–1950)

THE WORLD WARS

The Industrial Revolution in Europe accelerated the pace of innovations in the 19th century. The rapid technological advancement—new sources of power (steam and electricity), new means of transportation (railways, automobile, aircraft, steamships), faster means of communication (telegraph, telephone) and more effective medical interventions (vaccination, antiseptics) were invented—made Europe the colossal economic and military leader. Subsequently, rising economies like North America and Japan, and nationalist movements in Europe's colonies began to challenge Europe's dominance, leading to World War I.

World War I resulted in the Great Depression of the 1930s. As governments of the European countries struggled to find their way out of the economic dip, the US, the Soviet Union and Japan emerged as new economic and military powers, ending the dominance of Europe. With the power equation changing after World War I, one of the defining characteristics of this period was the rise of the US as the industrial powerhouse.

The 20th century, which has been sometime hailed as the American Century, saw the emergence of the US as the leading industrial, economic and military power during the period between the two world wars. There were several reasons why this has happened. The US was possibly the only big power which had remained relatively unaffected by World War I. The economies of Europe were still recovering from the huge costs of the four-year long devastating war. Another factor which contributed to the success of the US was the exploitation of natural resources of this continent-sized country as well as a massive increase in productivity, in both agriculture and industry. American entrepreneur were especially good at assimilating technological innovations and converting them into successful

Heralding the nuclear age

With the development and use of fission weapons in 1945, military warfare underwent a qualitative change. The post-World War II political scenario was defined by the cold war and competitive weapons development between the US and the Soviet Union. Bigger, more effective nuclear weapons and delivery systems were developed over the next few decades. Operation Ivy Mike in 1952, the first test of a fusion bomb by the US, generated a white mushroom cloud, shown here.

1913: Thermal cracking used for refining crude petroleum.

1914: World War I starts with wars between the major European powers but eventually engulfs the whole world.

1916: Einstein publishes his General Theory of Relativity.

1917: The October Revolution in Russia overthrows the Romanov dynasty and the establishment of the Soviet state. America enters the world war.

1918: World War I ends with the defeat of Germany. Weimar Republic established in Germany.

1920: In the aftermath of World War I, the great powers agree to create a League of Nations to keep peace. Women get the right to vote in the US.

1921: Founding of the Communist Party of China. Coup in Iran establishes Reza Khan as the king. Insulin discovered.

1922: Technicolor is introduced in movies.

1922: Benito Mussolini comes to power in Italy. Joyce publishes *Ulysses*. Ireland gains independence.

1923: Kemal Ataturk founds the Republic of Turkey after the fall of the Ottoman Empire. De Broglie proposes the wave-particle duality.

1925: Hitler publishes *Mein Kampf*.

1926: Schrödinger presents a self-consistent theory of quantum mechanics. Goddard flies the first liquid-fueled rocket. First movie with a sound track released.

1927: Lindberg's nonstop solo trans-Atlantic flight. Heisenberg formulates his Uncertainty Principle in quantum mechanics.

1928: Discovery of penicillin.

1929: Edwin Hubble discovers the expansion of the universe. Stock Markets crash leading to the Great Depression.

1930: Discovery of Pluto.

1931: The world's tallest building, The Empire State Building, built in New York.

1932: Establishment of the Kingdom of Saudi Arabia. Aldous Huxley publishes *Brave New World*.

1933: Nazis come to power in Germany and start a systematic campaign to exterminate the Jews in Germany. Start of the New Deal in the US. Roosevelt becomes president of the US.

1934: The Long March, massive military retreat of the Communist Army in China, begins.

1935: Mussolini captures Ethiopia.

1936: Start of the Spanish Civil War. Beginning of the Stalinist purges.

1937: Japanese forces attack China. Japanese occupation of China begins.

1938: Leaders of Britain, Germany, France and Italy sign the Munich Pact. Hitler annexes Austria. Volkswagen introduces the Beetle.

Volkswagen Beetle, introduced in Germany in 1938, was a bestseller

1939: Germany invades Poland, starting World War II. Russia and Germany sign a non-aggression pact. Russia invades Finland.

1940: Battle of Britain. Nazis invade Denmark, Belgium and France.

1941: Japanese aircraft attack Pearl Harbor. US enters the war. Atlantic Charter signed between the US and Britain. Germany attacks Soviet Union. Siege of Leningrad.

1942: Japan occupies Indonesia and Manila. Quit India movement started to end the British Rule

in India. Japanese forces attack Burma. German forces defeated at Stalingrad. The first self-sustaining controlled nuclear reaction initiated.

1943: Italy surrenders to the Allies. Teheran Conference between Roosevelt, Churchill and Stalin to plan the defeat of Hitler.

1944: Allied forces land in Normandy. London bombarded by Germany with V-2 rockets.

1945: Yalta conference to decide the fate of post-war Europe. Dresden bombings. Dropping of atomic bombs on Hiroshima and Nagasaki. World War II ends with the defeat of Japan and Germany. The development of nuclear weapons begins.

1946: First meeting of the United Nations. First computer is made.

1947: India gains independence. British India is partitioned into India and Pakistan. Polaroid camera and transistor are invented.

1948: Beginning of the Marshal Plan, a massive reconstruction effort for Europe. Independent state of Israel comes into being.

1949: NATO formed. People's Republic of China is born. Sukarno takes over as president of Indonesia. First civilian jet airplane takes off.

1950: India becomes a Republic. Beginning of the Korean War.

commercial enterprises. The massive industrial build up to feed the military machine for the world wars was left intact, and this too helped increase the economic growth of America.

In terms of inventions, what the world witnessed in the decades after World War I was unprecedented. The advances in all fields—agriculture, textiles, new materials, communications, transport, military technology, building technology and medicine—were so enormous that they impacted fundamentally the way human societies functioned. Apart from technological innovations and inventions, this was also the time when knowledge about nature expanded hugely. Quantum mechanics, the theory to explain natural phenomenon at submicroscopic scales, was one of the crowning achievements of the 1920s and 1930s. Einstein's General Theory of Relativity, an elegant description of how space and time behave at very large scales, was another intellectual *tour de force* of this period.

One other major change which happened in this period was that "little science" was replaced by "big science". In the previous centuries, science was usually done by individuals working in isolation while technological innovation came from craftsmen or men of leisure. This way of doing science underwent a radical change during the 20th century. Edison had already shown the way by establishing an industrial laboratory at Menlo Park toward the end of the 19th century. The 20th century witnessed the emergence of research universities (mostly in the US) as well as many industrial and government laboratories which served as the fountainhead of innovation in science and technology.

Manufacturing

Manufacturing, in the period spanning the two world wars, saw a major overhaul. The concept of the assembly line to improve efficiency in production, a brainchild of Henry Ford, became standard in many industries. Already, in 1911, the American engineer Frederick Taylor had enunciated his principle of scientific management, including extensive monitoring of the workers, dividing work into smaller units to be performed by individual workers, measurement of efficiency of workers and so on. Over the next few decades, American industrialists introduced these principles in manufacturing, and this improved productivity.

Agriculture

The introduction of agricultural machinery and chemical fertilizers, in the period before the 19th century, had already increased agricultural productivity immensely. Tractors and combined harvesters, machines which were hitherto using steam engines, started using internal combustion engines, leading to increased power and efficiency. The use of these machines in agriculture, especially in the larger farms of the US, was widespread.

In the war period, synthetic fertilizers, especially those based on petroleum, were introduced. The expansion of the chemical industry also led to the development of many chemical pesticides and herbicides. DDT (Dichloro-Dipheyl-Trichloroethylene) is one of the most well-known synthetic pesticides used extensively till the 1960s. Though it had been synthesized in the late 19th century, it was only in 1939 that Paul Müller, a Swiss chemist, discovered its efficacy against a wide range of pests. During World War II, it was used very effectively to reduce the mosquito menace in tropical regions; soon afterward it started being used in agriculture.

Textiles

Textiles, which were almost completely based on natural fibers like cotton, flax or silk, were now increasingly made with synthetic fibers. Though they had been known since the 19th century, the use of synthetic fibers became popular only in this period. Attempts to produce artificial silk had yielded a fiber which came to be known as rayon. This was the first artificial fiber to be used in textiles. The most popular synthetic fiber was, however, nylon, which was developed in 1930s at the DuPont Chemical Company. This fiber was synthesized by adopting some of the techniques which had been used to make plastics from petroleum products. Nylon was a very versatile fabric and, because of its light weight and resilience, found use in textiles, hosiery, parachutes and other products.

Ford assembly line

Though the assembly line was introduced in manufacturing in the 19th century, it was with Henry Ford's adaptation for car production that it became the mainstay of industry. In 1914, Ford installed the first conveyor belt assembly line in his Highland Park, Michigan car plant. This increased the efficiency and reduced the time and cost of manufacture, thereby making cars affordable. The war period also saw many innovations in management techniques with the introduction of time-motion studies to improve efficiency in production.

Petroleum

The popularity of the internal combustion engine, which was fed with gasoline, led to a greater dependence on petroleum. Petroleum refining techniques improved during this period, leading to the emergence of petrochemicals and plastics. Thermal cracking, discovered in 1913, was a technological breakthrough, used extensively during this period to refine crude petroleum. In this process, the heavier molecules present in the petroleum were "cracked" into lighter ones like petrol by employing heat and pressure. This was replaced in 1936 by catalytic cracking, a process using a catalyst—namely zeolite—to break up petroleum fractions. The catalyst speeded up the cracking reaction, and led to an increased yield of petrol.

Plastic and Other Materials

The introduction of plastics was one of the biggest inventions of this period. It was a result of the improvements in petroleum refining and a better understanding of the chemistry of petroleum. Plastics had been made from organic chemicals in the 19th century—such as celluloid and Parkesine—but they were neither manufactured in any quantity nor did they become popular. In 1908, Leo Baekeland, a Belgium-born chemist-entrepreneur, patented a new material called Bakelite. Made from formaldehyde and other chemicals, Bakelite had insulating properties that made it useful in electrical appliances.

After the discovery of thermal cracking and later catalytic cracking, the plastics industry really boomed. Many new plastics, mostly made from petroleum, were introduced for a variety of uses. In paints, pipes, films and a host of other applications, plastics replaced other materials.

They were also used extensively in the building and construction industry.

Though plastic was the wonder material of this period, other important materials were introduced too. Stainless steel was first made in 1913 by alloying steel with chromium. Soon, alloys of steel—itself made with iron and carbon—with various metals in differing compositions were developed. The metallurgy industry became more specialized with special steels being made for specific purposes—for use in electrical wires, internal combustion engines, jet engines and utensils.

Aluminum had been discovered and isolated in the 19th century, but it could not be produced in large quantities till the process of extraction using electricity was discovered in 1886. With the huge increase in electricity generation in the early 20th century, the extraction of aluminum became feasible. Aluminum was a lightweight alternative to steel and copper in many applications. It found use in the electrical industry, buildings, aircraft construction and household appliances.

During World War I, the supplies of natural rubber from South America and the Far East were disrupted. This led to the development of an industrial process to make synthetic rubber by the German scientists, though at least one form of synthetic rubber had been made in the 1890s in the laboratory. The Germans experimented with making rubber from butadiene, a petroleum byproduct. After the war, the use of synthetic rubber declined, though scientific investigations on the process to make different rubber-like materials continued. In 1933, the Germans invented a new form of rubber called Buna S, in which butadiene linked in a chain with styrene; this rubber was a huge commercial success.

Petroleum refining

As the demand for petroleum increased in the 20th century, petroleum-refining techniques also evolved. The discovery of thermal cracking in 1913 marked the real beginning of the petrochemical industry. With thermal cracking, and later catalytic cracking, it became possible to derive the useful components of crude oil on a large scale. The plastic industry was also a direct offshoot of this.

Electricity

While steam power had been the fuel of the Industrial Revolution as well as of the 19th century, electricity was the major source of power in the 20th century. Electrical generation and its use in motors and other appliances were well known by 1914. However, there was a huge increase in the production of electricity in the early 20th century as more and more households in the US and Europe started using it. In manufacturing and other industries, electric power and the internal combustion engine replaced steam almost totally. Though initially direct current was used in most places, it was soon realized that alternating current was much easier to transport over large distances without any significant loss.

Generation of electricity was mostly using steam turbines, which burnt coal or oil, or by using the energy of falling or running water—as in the case of the pioneering hydroelectric plant at the Niagara Falls. Huge thermal power plants were set up by the 1930s as the consumption of electricity grew. Also, transmission lines to deliver electricity everywhere were installed with substations at regular intervals.

Atom Bomb

Another invention of this period was the unleashing of the energy in the atomic nucleus. The theoretical work for using the atomic nucleus as a source of unimaginable power had been going on throughout the 1920s and the 1930s, but this was rarely with any technological purpose. The atomic nucleus could be understood much better with the formulation of quantum mechanics and with research in nuclear physics. In 1932, the neutron, a sub-atomic constituent of the nucleus, was discovered, and several scientists around the world started experimenting with the behavior of matter when being bombarded with neutrons. Among them were Otto Hahn and his collaborators in Germany, who in 1939 showed that heavier atomic nuclei could be split apart using particular kinds of neutrons. Another Italian scientist, Enrico Fermi, had carried out several experiments with radioactive uranium and neutrons, too.

In 1938, Fermi came to the US to escape fascism in Italy. In 1939, he started working on the Manhattan Project to construct an atomic weapon. This US initiative was in response to the fear that Germany might be developing an atomic bomb. Fermi was assigned the responsibility of generating a controlled, self-sustaining nuclear reaction; in 1942, the first such reaction took place in an atomic pile at the University of Chicago. The atomic genie was now out of the bottle, and this would change the politics and history of the world over the next few decades.

Transportation

The invention of the internal combustion engine and airplane had already revolutionized transportation in the period before World War I. However, both these technologies saw massive improvements in the period between the two wars. Automobiles became more and more common, especially in the US after the introduction of assembly-line production by Henry Ford. There were many innovations which improved the quality and durability of automobiles. These included a self-starter (to do away with the

need for cranking to start the engine), improved suspensions and four-wheel brakes which used a hydraulic system and synchronized gears.

Innovations in flying technology had already started after the first flight by the Wright Brothers. But the world war proved to be a major hindrance for any further development. After the war, better and longer flying airplanes were introduced—a non-stop trans-Atlantic flight was accomplished in 1919. One of the major developments in the early history of airplanes was the introduction of multiple engines. A single engine, with its limited power, could not provide enough lift for carrying fuel, passengers and cargo.

In 1913, the Russian-born American aviation pioneer Igor Sikorsky built the first four-engine plane. But it took many years before a four-engine plane could enter commercial aviation. In 1935, the first flight across the Pacific—from California to Manila—was accomplished with a four-engine plane. With this, the era of commercial aviation had truly begun. The beginning of World War II provided another incentive for the development of better, faster and more powerful airplanes. Several different airplanes—such as those which could

Making of the atom bomb

With the discovery of radioactivity and later of nuclear fission, there was an increased interest in understanding the nature of matter at the submicroscopic level. On the theoretical front, the advances in quantum mechanics were critical to this endeavor. On the experimental front, machines like the cyclotron proved invaluable to enhance knowledge about the nature of matter. Here Dr E.O. Lawrence is working on the cyclotron-producer atom-smashing machine in the development of the atomic bomb.

Sikorsky and his helicopter

Drawings of a helicopter-like machine are found in the notebooks of the Renaissance polymath Leonardo da Vinci. Though various attempts were made to construct helicopters in the 19th and early 20th century, none of them was completely successful. Igor Sikorsky, a Russian-born American aeronautical engineer, after a long period of development, made several successful test-flights with his helicopter in 1939–40.

V-2 rocket

First used in war in 1944, the Germans developed the V-2 rocket using the rocket technology of R.H. Goddard. Experimenting with solid-propellant rockets first, Goddard tested the first successful liquid-propelled rocket engine in 1926. V-2 was the first ballistic missile which could fly at supersonic speed at an altitude of 50 miles (80 km).

serve as army troop transport—were designed and used during the war.

The jet engine was another revolutionary development of this period. Although the idea of using jet propulsion was an old one, it was only in the 1920s that serious work began in this area. In 1930, a British aircraft apprentice, Frank Whittle, filed a patent for a turbojet propulsion system. It took seven years for an engine of this design to be made and tested. In the meantime, engineers in Germany were also working on a jet engine, and in 1939, the He 178, the aircraft fitted with a jet engine, took its first flight in a private venture by the Heinkel company. Subsequently, several planes with jet engines were used during World War II. In 1949, the first civilian jet airplane made its maiden voyage.

After the success of the Wright brothers with their biplane, many developments took place in the airplane technology. However, there was another flying machine which also emerged during this period: the helicopter, an aircraft being able to take off vertically and to stand stationary in air. Leonardo da Vinci had already sketched the essential elements of a helicopter but translating this into a practical design had to wait for the development of appropriate power sources. Several engineers, in many countries over a period of time, made incremental changes in design to make a successful helicopter. In 1912, contra-rotating rotors and pitch control were invented by the Danish inventor Jacob Christian Ellehammer. Many engineers in Germany worked on helicopter

design and, in 1936, produced a three-rotor model. Three years later, the Russian–American aircraft engineer Igor Sikorsky built a helicopter with a single main rotor and a vertical tail rotor. With its first successful flight in 1939, VS-300 was the prototype single-rotor helicopter. The Sikorsky design was very successful and most subsequent developments were modifications or this basic design.

Rockets

While airplanes and flying machines were in the limelight during this period, another technology saw some major developments during this time. Rockets, used by the Chinese since the turn of the millennium, gained a renewed interest, partly because of the fascination with space travel, mentioned by Jules Verne in his hugely popular books like *From the Earth to the Moon*. Konstantin Tsiolkovsky, a Russian scientist, worked out the theoretical foundations of practical rocket propulsion and motion around the turn of the century.

However, the major innovations came with the research of the American inventor Robert Goddard, who worked on many problems associated with rocket flight and propulsion. His work included designs of rockets with solid and liquid fuel. In 1926, the first liquid-fueled rocket was test-fired in Auburn, Massachusetts. The rocket was in the air for only 2.5 seconds, attaining an altitude of 41 ft (12.5 m). Goddard, called the Father of Modern Rocketry, continued to work on improving the designs of his rockets throughout his life, though he could not find enough resources to build most of them. In 1919, Goddard published the essentials of rocketry in a pamphlet titled "A Method of Reaching Extreme Altitudes". His work was extensively used by the Germans who built the infamous V-2 rockets which were used with devastating effect during the latter part of World War II.

Medicine

The 19th century had seen a dramatic change in the theory and practice of medicine. The growing acceptance of bacteriology and the germ theory meant that illness came to be seen as something which had material origins and could thus be amenable to treatment accordingly. The discovery of various disease-causing microbes and vaccines meant that some of the major killer diseases were not as devastating in terms of loss of human life as before. The use of antiseptics was also important in controlling infections, especially surgical infections.

With the scientific theory of medicine, the stage was set for the big breakthroughs of medicine in the 20th century, especially pharmacology. X-rays, previously used for scientific curiosity, began to be used widely for diagnostic purposes, though the harmful effects of extended exposure to this deadly radiation were not properly appreciated till much later.

World War I saw the improvement of many surgical techniques to treat the wounded. Nevertheless, millions perished because of lack of an effective treatment for wound infections. The quest for finding an antibacterial agent was undertaken by several scientists. One of them was a British bacteriologist, Alexander Fleming. In 1922, he discovered that lysozyme, a chemical contained in human tears, had some antibacterial properties. Further research showed that this was ineffective against most infections and also killed many non-disease causing bacteria.

In 1928, purely by accident, Fleming, during the course of his experiments, discovered that the presence of a particular mold did not allow bacterial colonies to grow. The mold, which belonged to the penicillium family of fungus, was thus the first antibiotic. Fleming extracted the antibacterial chemical from the mold and demonstrated its effect on various kinds of bacteria. However, he did not succeed in producing and refining enough of the substance, and his work was soon forgotten.

In 1938, two chemists from Oxford University, Howard Florey and Ernst Chain, revisited Fleming's work, and using some clever techniques, managed to purify the chemical and test it on animals and humans. Scientists in various parts of the world were also experimenting and discovering different antibiotics which worked against specific microbes, like the bacteria causing tuberculosis or the microbe causing strep throat. The world now knew that finally an effective agent against bacteria had been found. The problem though was still to manufacture antibiotics in sufficient quantities to be able to make a difference in public health.

The necessary impetus came with World War II when the British and American governments realized the importance of making antibiotics in sufficient quantities to be used on the wounded. By 1942, antibiotics were being manufactured industrially, and their efficacy in preventing millions of fatalities during the war was obvious.

This period was also the time when many vitamins were identified and their necessity for the human body was understood. Deficiency diseases like scurvy and beriberi had been known for a long time and, over the years, specific preventive agents had been discovered. For instance, it had been known for more than a century that eating

Alexander Fleming in his laboratory

The Scottish bacteriologist Sir Alexander Fleming revolutionized the treatment of bacterial infections by discovering penicillin, the first antibiotic. Though Fleming discovered the wonder drug in 1929, penicillin could not be produced in large quantities for more than a decade. As casualties mounted during World War II, a team of scientists working on the drug rushed to set up a network of mini factories for penicillin production and spurred large-scale production by US pharmaceutical companies, saving millions of lives in the process.

citrus fruits prevents scurvy. But the specific agents whose deficiency caused the disease had not been identified. In 1912, a Polish chemist, Casimir Funk, identified one of the complex of chemicals which prevented some deficiencies. He called this complex Vitamine. Over the next few years, many scientists worked to discover the chemicals responsible for preventing the diseases and these were all given the generic name Vitamin, with the specific chemical being identified with a suffix. By 1943, almost all the known vitamins had been identified and their importance for the human body understood.

In 1922, the first insulin injection was given to a diabetic with stupendous results. Frederick Banting, a Canadian doctor, had managed to extract the chemical from calf pancreas. Subsequently, insulin extracted from animals became the standard treatment for diabetes.

In diagnostics, X-rays were already being used. In 1920, the German physiologist Horst Berger began studying the electrical phenomenon in the human brain. He was using a device which had been used earlier but had never been named.

Berger carried out extensive studies in recording the electrical activity of the brain and named the device Electroencephalograph or EEG. Over the years, this device underwent many modifications and is still used for researching the physiology and function of the brain.

Communications and Media

With the development of the radio, the telegraph, the telephone, photography and the phonograph, the 19th century was truly the golden period of communications. The impact of these inventions was truly stupendous in bridging distances and time. In comparison, the period between the two wars was not that fruitful.

The idea of being able to transmit images in combination with sound, as in the radio, was conceived in the early part of the 20th century. There were several attempts to realize the idea but none were successful. In 1907, Boris Rosing at St Petersburg worked out the details of how such a device could be constructed using the cathode ray tube which had been invented in the last century. However, his ideas never took practical shape.

In 1923, one of Rosing's students, Vladimir Zworykin, working in the US, filed a patent for a television system which utilized a cathode ray tube. After many years of hard work and a lot of money spent by his employer RCA, Zworykin produced the first working television system in 1931. There was a patent dispute which finally got settled in 1939 and, later that year, the first public broadcast of the New York World Fair took place.

There were many improvements in the design of the telephone, the telephone exchanges and transmission systems as well as radio receivers and transmissions during this time. Radio became an important medium for news and entertainment as well as propaganda. However, the most important invention for radio and television came in 1947. This was the transistor invented at the Bell Labs by a team comprising William Shockley, John Bardeen and Walter Brattain. Though there had been research in finding a more efficient and portable alternative to the bulky electron tubes which were used in radio and television for a long time, it was only after the onset of World War II that a lot of money was provided for radar research. The invention of transistor—a miniature device which used semiconductors like silicon and germanium—triggered many innovations in electronics. After the initial fabrication of the successful device in 1947, different kinds of transistors were developed at the Bell Labs over the next few years.

Radio technology also found use in the military in the development of the radar system. Ever since the discovery of radio waves, people had toyed with the idea of using these waves for navigation and the detection of obstacles. Prior to the war, there were some radar systems which were being used by the militaries in various countries. However, it was during the war that the real development took place—the key invention was the cavity magnetron oscillator developed by British scientists in 1939. This device then helped construct a microwave radar at MIT in 1940, which was used with great efficacy during the world war.

By 1914, motion pictures had become popular as a source of documentation as well as entertainment. These were silent movies since the problems associated with synchronizing sound with the images as well as amplifying the sound enough to be heard by the audience were not solved. These problems got solved in the 1920s leading to the invention of talkies. Technicolor was also introduced in movies around the same. With color and sound, the whole experience of movie watching was revolutionized.

Charlie Chaplin's *The Great Dictator*

Motion pictures became a major industry after the introduction of sound and color in 1920s. It was around the same time that the studio system got established in the Hollywood movie industry. Al Jolson's *Jazz Singer*, released in 1927, marked the beginning of talkies. But Charlie Chaplin, whose Tramp character was famous for his pantomime, was hesitant to make the transition to talkies. His first talkie, *The Great Dictator*, came in 1940 and was well received.

Computers

One of the major advances during the period between 1914 and 1950 was the development of computers. Computers had a long history as calculating machines—the abacus, Babbage's difference engine, the slide rule and finally Hollerith's census machines were all mechanical devices which could help in doing simple arithmetic operations like addition, subtraction and multiplication. However, it was only after the development of the first general-purpose electronic computer that modern computers came into being.

In 1930, Vannevar Bush at MIT developed a differential analyzer, which was a mechanical machine to solve specific problems. This was an analog machine but proved to be very useful, especially in the aerospace industry. Patented by Bush in 1935, the device used to calculate the ballistic table in World War II.

Beginning in 1937, another inventor, Howard Aiken, developed several devices which used electromagnetic relays and vacuum tubes. His machines were huge and consisted of many mechanical parts. While this work was going on in the US, a mathematics professor in England, Alan Turing, was formulating the theoretical ideas behind a universal computer.

The world war provided the necessary push to research in computing machines. Computing power was needed to make artillery tables, for breaking codes and other war-related activities. In 1943, a machine called Colossus was made in the UK to crack the German ciphers while another machine, called the Z4, was developed in Germany.

However, it was in the US that the first general programmable electronic digital calculating machine ENIAC (Electronic Numerical Integrator And Computer) was built in 1946. A powerful calculating machine, the ENIAC was used for calculating artillery tables for Ballistic Research Laboratory of the US Army. The brain behind the machine, John von Neumann, a Hungarian émigré, was responsible for the concept of a stored program in a computer. This was a major advance, and in the next few years, several machines using this idea were built in the UK and in the US.

With the essential theoretical ideas of a programmable digital machine in place, the next few decades saw the development of many machines and programming languages. The invention of the transistor, and later the integrated circuit, made these devices possible.

Enabling Inventions

The period thus between the beginning of World War I and 1950 saw the emergence of several key technologies like television, antibiotics, liquid-fueled rockets, plastics and the jet engine.

However, more importantly, this was the period when much work was done in increasing the understanding of nature. Quantum mechanics was by now the well-established theory to understand the submicroscopic domain. It had developed from the initial ideas of Max Planck, Einstein and Niels Bohr among others. By the 1930s, the essential ideas and foundations of the quantum theory were well established.

Another major scientific achievement of this period enhanced our understanding of the universe. Einstein's General Theory of Relativity provided the theoretical framework for this. In the 1920s, with the work of astronomers like Fritz Zwicky, Edwin Hubble and others, the ideas of an expanding universe were established. At the same time, there was increased knowledge about the other constituents of the universe.

During the period between the world wars the US had become the major economic and scientific power. This was further reinforced after World War II when the US furthered its lead in the economic, scientific and technological spheres. The only real scientific competition was provided by the Soviet Union, which remained as the second major superpower after World War II. The next few decades were dominated by this competition between the two powers for scientific and military supremacy, which would not only include atomic weapons but also the first steps into space.

A panel of IBM's ENIAC computer

In 1946, ENIAC, a powerful computing machine, was commissioned. This behemoth was not very advanced technologically and had many limitations; nevertheless it was a huge improvement over its predecessors, executing over 5,000 additions per second. It occupied a 50-by-30 ft (15-by-9 m) basement and had thousands of vacuum tubes, capacitors and relays. Although designed to help in the war effort, it was completed after the war was over. The first calculations on the ENIAC were for the design of the hydrogen bomb.

R. Oppenheimer and How Inventions Shape History

J. Robert Oppenheimer

The brilliant physicist Robert Oppenheimer headed the Manhattan Project to develop the first atomic bomb. The Manhattan project was a huge organizational and scientific challenge which involved coordinating the work of hundreds of scientists and thousands of engineers, spread over many locations.

The 1930s was an eventful decade in the history of physics. In 1934, Irène and Frédéric Joliot-Curie had discovered that radioactivity, the spontaneous emission of particles by certain heavy elements, could also be induced artificially using alpha particles. Their work was followed up by the Italian scientist Enrico Fermi who for the first time showed that neutrons could be used for this purpose. In 1938, German scientists Otto Hahn and Fritz Strassmann showed that bombarding uranium with neutrons produced barium. Their work was interpreted by Otto Frisch as splitting of the uranium nucleus by the neutron into smaller constituents with the release of energy. Nuclear fission, as this process is known, was thus shown to be possible.

The importance of this development has been observed by the scientists. The rise of Nazism had forced many of the scientists from Europe to emigrate to the US and Britain to escape persecution. By 1938, it was clear to most scientists that the Nazis were actively carrying out research to realize the destructive potential of nuclear fission. Some of them, including Einstein, wrote to President Roosevelt about such a possibility and the devastating consequences of a Nazi fission bomb. The Manhattan Project, a massive enterprise to construct an atomic weapon before the Nazis—employing over 100,000 people at more than 30 locations—was a result of these developments.

The scientific director of the project was a brilliant theoretical physicist, J. Robert Oppenheimer. Oppenheimer coordinated the work of a virtual who's who of scientists over a period of almost five years to produce the first atomic or nuclear fission bomb. Although the basic principles of such a weapon were clear, the design and technology of a working atomic bomb posed a huge intellectual and administrative challenge.

After many false starts, in the summer of 1945, two types of bombs were finally completed—uranium and plutonium based weapons. The uranium bomb could not be tested since there was just about enough uranium available for one weapon. Thus, early morning on July 16, 1945, at a remote firing range in New Mexico, the plutonium bomb, named Trinity, was tested. The world would never be the same again after this one event.

The test was a stupendous success scientifically. The explosion was unlike anything ever witnessed by humans. The bomb was equivalent to 20 kilotons of TNT and it left a crater 10 ft (3 m) deep and more than a 1,000 ft (300 m) wide in the desert. The desert, for a few seconds, was illuminated in a light brighter than sunlight and the mushroom cloud loomed almost 7.5 miles (12 km) into the atmosphere. The enormity of the event led Oppenheimer, an erudite scholar, to recollect lines from the Hindu scripture Bhagvad Gita, "now I have become Time, the destroyer of worlds".

Within a month of the test, the whole world witnessed the destructive power of the new weapons when the US dropped atomic bombs on the Japanese cities of Hiroshima and Nagasaki. The devastation caused to life and property was unprecedented—the two cities were completely razed to the ground and hundreds of thousands of people died immediately. The deaths and illness caused by radiation continued for decades.

The invention of the nuclear weapon and its use could be regarded as one of the defining moments of the 20th century. Europe had been divided up into spheres of influence between the US and the Soviet Union. The American bomb led the Soviets to double their efforts to produce their own weapon which was tested in 1949. By then, the US was on its way to produce a much more powerful fusion weapon. In 1952, the first American fusion weapon was tested, followed a few months later by the Soviet one. The nuclear race was now under way between the two blocs; history and politics in the latter half of the 20th century would be molded by this arms race.

Scientific and technological inventions have always been a major determining factor in shaping history of mankind. In this light, perhaps the biggest invention in history was the domestication of plants during the Neolithic period leading to agriculture, which necessitated settled communities and finally civilization based on the surplus availability of food. On a much smaller scale, the use of metals, the invention of the wheel, the development writing, the domestication of the horse, and after the invention of gunpowder, printing, steam engine, telegraph, telephone and radio could be considered key determinants at various points in human history. These and other such inventions have had a profound impact on the way humans have interacted with each other or with nature. This has further led to major changes in history, society and politics. For instance, the domestication of the horse followed by the stirrup in the central Asian steppes gave the nomadic tribes a distinct advantage in warfare and was a major factor in the Mongol conquests. The invention of printing in the 15th century played a major role in democratization of knowledge which had hitherto been the monopoly of the clergy. The distribution of printed material led to the Reformation of the Church. Similarly, the steam engine made possible the Industrial Revolution, while inventions such as the telegraph and telephone bridged distances which were previously unimaginable.

Though inventions like those mentioned above and others have played a role in determining historical events, it has not been the case that these have operated in isolation. The importance of any innovation or invention is ultimately determined by external factors. The receptiveness of the society to the innovation determines its use. Social structures are also important in the dissemination of any innovation. In many ancient civilizations, to cite an example, an invention like writing could not permeate beyond the royal and the priestly classes. On the other hand, sometimes society itself is shaped by technologies—for instance, the increased urbanization and the movement

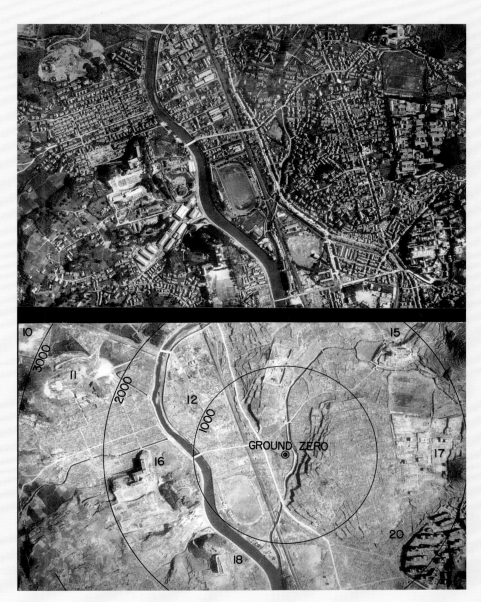

of the peasantry to the factories caused by the Industrial Revolution.

Thus the two-way relationship between technology and society is a historical truth. In the case of atomic weapons, it is undeniable that this one technology formed the underlying motif of the cold war which the world saw in the latter part of the 20th century. On the other hand, one of the primary reasons for the development of atomic weapons, that is the technological realization of the theoretical ideas, was the political reality during the world war with the rise of fascism.

The danger posed by nuclear weapons to humanity was recognized by many scientists and public figures soon after the Hiroshima and Nagasaki disasters. This led to several public movements which were ultimately responsible for various treaties like the Comprehensive Test Ban Treaty (CTBT) and the Nuclear Non-Proliferation Treaty (NPT). However, nuclear weapons are now recognized as something that humanity has learnt to live with, though it is a unique weapon in history whose efficacy is based on the threat of its use.

Nagasaki, before and after the bombing

On August 9, 1945, the second atomic bomb was dropped on the ship-building city of Nagasaki in Japan. The mushroom cloud rose over 60,000 ft (18,288 m) in the air and everything in a radius of about 2 miles (3.2 km) was pulverized. The world had never seen destruction of this kind in history. The aerial photographs focus on the area of the city where the bomb was dropped (ground zero) before and after the bombing, showing the scale of devastation.

Anti-nuclear protests

Soon after World War II and the devastation at Hiroshima and Nagasaki, several scientists and policy makers realized the dangers of nuclear weapons. Throughout the 1950s and 1960s, during the cold war era, the anti-nuclear movement grew, but it was only during the 1970s that it became a global movement. The photograph shows a South Korean protestor demonstrating in Seoul against the nuclear weapons program of North Korea.

POST-WAR 20TH CENTURY
DISCOVERING OUTER SPACE (1950–2000)

Man on the moon

On July 20, 1969, Neil Armstrong, the Apollo 11 astronaut, became the first human to walk on the moon. He was soon joined by Edwin "Buzz" Aldrin, and the two collected samples, took photographs and planted the American flag on the lunar surface. Here Aldrin, deploying scientific experiments on the moon, is photographed by Armstrong. During the next three years, six spacecraft and 12 astronauts landed on our satellite.

RIVALRY OF THE TWO SUPERPOWERS

The second-half of the 20th century was an eventful period in world history. While on the one hand, this was a period of reconstruction after the devastation wrought by World War II, this was also a time when the cold war, a period of intense rivalry between the Soviets and the Americans, was at its peak. The cold war was not just restricted to the military and political realm but extended into

the technological domain as well. In fact, till the breakup of the Soviet Union in 1989, the cold war was the defining framework within which all major developments were viewed on either side.

The aftermath of the world war left almost total devastation in Europe, Japan and, to an extent, in the Soviet Union. The political leadership of the US and the Soviet Union had agreed to a division of Europe into spheres of influence, and thus, in the post-war period, countries in Europe were mostly part of either the Eastern Bloc or the Western

1950: War starts in the Korean peninsula. Jordan annexes West Bank after the Israel-Arab war.

1951: Australia and New Zealand join the US-led alliance ANZUS.

1952: Mau Mau rebellion starts in Kenya against the British rule. Batista becomes dictator of Cuba. Jonas Salk develops the polio vaccine. US detonates the first hydrogen bomb.

1953: Korea is divided into North Korea and South Korea. Josef Stalin dies in Russia. Edmund Hillary and Tenzing Norgay scale the Mount Everest. James Watson and Francis Crick decipher the structure of DNA.

1954: Vietnam is partitioned. US supports the military to topple the government in Guatemala. The solar cell is invented by Daryl Chaplin, Calvin Fuller and Gerald Pearson.

1955: Bandung conference in Indonesia to form a group of non-aligned nations. Optic fiber invented.

1956: Nasser nationalizes the Suez Canal, leading to a war with Britain, France and Israel. Popular uprising in Hungary. Christopher Cockerel invents the hovercraft.

1957: Treaty of Rome establishes the European Economic Community. Hundred Flowers Movement in China. Soviet Union launches the Sputnik. FORTRAN is launched.

1958: NASA established. Gordon Gould invents the laser. The integrated circuit is developed by Jack Kilby and Robert Noyce.

1959: China crushes a rebellion in Tibet and the Dalai Lama flees to India. Luna 2, a Soviet probe, reaches the moon. The internal pacemaker is invented by Wilson Greatbach.

1960: Seventeen African countries gain independence from colonial rule.

1961: Berlin wall is built. Yuri Gagarin becomes the first human in space.

1962: Indo–China war over border dispute. US launches Mariner 2 to Venus. US and USSR come close to a nuclear conflict over Soviet missiles in Cuba. The audio cassette is invented.

1963: John F. Kennedy assassinated. The US, USSR and Britain sign the limited test ban treaty. The first videodisc invented.

1964: Civil Rights Act passed in the US. Military coup in Brazil. Acrylic paint is invented. BASIC is developed.

1965: Escalation of US war in Vietnam. Intelsat 1, the first geostationary communications satellite, launched. The CD is invented by James Russell.

1966: Cultural Revolution in China.

1967: Six-day War between Arabs and Israel. The first hand-held calculator is invented.

1968: Martin Luther King assassinated. The first heart transplant carried out in humans by Christiaan Bernard in South Africa. The first computer with integrated circuits is made.

1969: Neil Armstrong lands on the moon. Arpanet, the precursor to the Internet, established. ATM and bar-code scanners are invented.

1970: US invades Cambodia. The floppy disk invented by Alan Shugart.

1971: Bangladesh gets independence from Pakistan. The Soviet space station Salyut I is established. Intel 4004, the first microprocessor, released. The dot-matrix printer, VCR and LCD invented.

1973: Arab-Israel War leads to an oil embargo. Pinochet leads a military coup in Chile to topple the socialist government. Gene splicing

begins. Ethernet is invented by Robert Metcalfe.

1974: Richard Nixon resigns over Watergate scandal. Emperor Haile Selassie overthrown in Ethiopia. Liposuction invented.

1975: Helsinki accords signed to defuse cold war in Europe. The first personal computer launched and the laser printer invented.

1976: Soweto uprising in South Africa. Viking 1 and 2 land on Mars. The ink-jet printer invented.

1977: Brezhnev comes to power in USSR. Magnetic resonance imaging invented by Raymond V. Damadian.

1978: Soviets invade Afghanistan. John Paul becomes Pope. Camp David accord signed between Egypt and Israel. The first test-tube baby born.

A mobile phone and a palm-top

1979: Three Mile Island accident in the US. Islamic revolution in Iran. Margaret Thatcher becomes prime minister in the UK. Cellular phones and walkman invented.

1980: Iran-Iraq war starts. Mugabe leads the independence struggle in

Zimbabwe. Hepatitis-B vaccine is developed.

1981: US launches the first reusable space shuttle Columbia. Anwar Sadat assassinated in Cairo. IBM PC and MS-DOS invented.

1982: Israel invades Lebanon. UK–Argentina clash over the Falklands Islands.

1983: US invades Grenada. The AIDS virus identified. Apple Lisa invented.

1984: Leak of poisonous gas at a chemical plant in Bhopal, India. CD-ROM invented.

1985: Windows invented by Microsoft.

1986: Chernobyl nuclear reactor explodes. Glasnost begins in the USSR. Space shuttle Challenger explodes and the USSR launches Mir. Super conductor invented.

1989: Fall of the Berlin Wall.

1990: Nelson Mandela released from prison. NASA launches the Hubble Space Telescope. World Wide Web is created.

1991: Boris Yeltsin elected president in USSR. The first Gulf War between Iraq and the US. Collapse of the Soviet Union.

1993: Middle East peace accord signed between the Palestinians and Israelis. The first web browser, Mosaic, developed in the US. Pentium processor invented.

1994: Mandela elected President of the Republic of South Africa.

1995: DVD invented.

1997: Hong Kong transferred back to China. Cloning of the sheep Dolly. NASA spacecraft lands on Mars.

1999: Macau falls back to China.

loc. Though this division was more pronounced Europe, it actually extended throughout the hole world.

Militarily, this was the period which saw several ng, drawn-out conflicts. Countries of Asia and frica got their freedom from colonial rule— ostly, freedom was attained by waging liberation ars. In 1950, the Korean War—the first major ost-war conflict which pitted the two blocs opposing sides—began, which was to last for ree years. It was the first time that a war was eing fought in the world under the shadow of a uclear holocaust.

The 1950s also witnessed the witch-hunting progressive elements from public life, the cademia as well as the entertainment industry the US. Anti-communism became the dominant eology in the US and any sign of protest against e established order was suppressed. This was so the time of an unprecedented economic boom America, and after almost three decades, the merican economy became what it was like just efore the Great Depression.

The arms race continued throughout this period ith each side developing bigger and more effective comic bombs and delivery systems. The 1950s was so the time when human beings reached out into ace and opened up a new frontier. The launching the Sputnik by the Soviets was followed by e developments in space travel in the US and e Soviet Union, which were among the biggest chnological achievements of the period.

Technologically, the latter half of the 20th ntury was a very productive period. Technologies hich define the industrial, or in fact the post-dustrial, era were introduced during this time. pace travel, microelectronics, computers, ternet, high-yielding varieties of cereals, nuclear ower and digital entertainment were some xamples. Some of these innovations, especially electronics, automation and computing, were volutionary enough to change the basic paradigm industrialization and consumption. Agriculture ad already been replaced by manufacturing as the ominant element in economic activity before this eriod in the West. This period saw unprecedented rowth of the service industry, which became the ost important economic agent in the West.

However, the massive increase in consumption the West as well as the economic growth in e developing countries were at the cost of normous environmental damage, a fact which ecame increasingly evident toward the end of e 20th century. This led to the development of nvironmental-friendly technologies to conserve e depleting natural resources.

Exploration the Outer Space

Designated the International Geophysical Year, 1957 witnessed the first attempts by humans to explore the outer space. In fact two years earlier, the US had announced to launch an artificial satellite in this year. However, on October 4, 1957, the Soviets revealed that they had successfully launched a satellite to orbit the earth. Sputnik I, as the satellite was called, was launched using a modified ICBM rocket-launcher and circled the earth for about three months before burning out in the atmosphere on re-entry.

The launch of the Sputnik was a complete shock for the West as it had always been assumed that

the Soviets did not possess the technological wherewithal needed for such a complex operation. Sputnik I was followed by the launch in November of an even larger satellite, Sputnik II, carrying the dog Laika into space. To make matters worse, the American effort to launch a satellite using the Vanguard rocket failed miserably when the rocket exploded on the launch pad in December 1957. It was only on January 31, 1958, that the first American satellite, Explorer 1, was successfully placed in orbit. With these events, the space race had truly begun.

The cold war ideology saw the Soviet Union's triumph in space as a defeat for the West. A massive program to encourage science and technology education and research at all levels was initiated in the US. For space exploration, a separate agency, NASA, was formed in 1958, which rapidly grew into a huge organization.

The West accepted another punch when in 1961, Yuri Gagarin, a Soviet test pilot, was launched

Sputnik III

The Sputnik program included a series of unmanned spacecraft launched by the Soviet Union for space exploration. Sputnik I was the first man-made object to be put into orbit around the earth, on October 4, 1957. Sputnik II, launched in the same year, carried the first living being, a dog named Laika, into space. The photograph shows Sputnik III, launched on May 15, 1958, on display at Soviet exhibit.

Yuri Gagarin

The Soviet Union was the first to send a human being into space. Yuri Gagarin, a Soviet cosmonaut, went into space aboard Vostok I on April 12, 1961. Being the first human to enter space and orbit the earth, Gagarin became a celebrity throughout the world and received medals and honors from many countries. Gagarin's space flight proved that it was possible for humans to sojourn in space and thus triggered a series of attempts to send man into space, including the Apollo mission.

Apollo 11 spacecraft take-off

The enormous power needed for carrying astronauts to the moon was provided by Saturn V boosters. Saturn V consisted of three stages and an instrument unit which contained the guidance instruments. It was the most powerful rocket ever developed and its engines could take the spacecraft to a speed of over 15,000 mph (24,139 kmph). The photograph is a composite photograph of the lift-off of the Apollo 11 spacecraft from Cape Kennedy.

successfully into space, staying there for about two hours before parachuting back to the ground. The humans had finally taken their first step in outer space. As a reaction to this, President Kennedy announced that Americans would land on the moon before 1970—an unbelievably ambitious goal, given the primitive nature of technology and the challenges involved.

The next few years saw accelerated development of the systems needed to put man on the moon—launch systems powerful enough to take the increased payload, lunar orbiter, lunar lander and all the associated communication and power systems. The Apollo program went through many phases and finally, at the end of 1968, Apollo 8 became the first-manned spacecraft to orbit the moon and successfully return to the earth.

On July 20, 1969, Neil Armstrong landed on the moon, an event which marked, as he eloquently put it, "a small step for a man, a giant leap for mankind." This first landing of humans on another celestial body had a huge political significance. The Americans had finally achieved an advance in space exploration.

Throughout the 1960s, though the focus laid on manned space flight and human landing on the moon, both the US and the Soviet Union had been sending spacecraft to explore other planets and objects in the solar system. The Viking, the Mariner and the Soviet Venera missions were launched to gather information from Mars and Venus. While the spacecraft Mariner made fly-bys of the innermost planet Mercury in 1974–75, the Viking landed on Mars in 1976—this was an important milestone in space exploration.

In the last quarter of the century, probes were launched to explore other planets and even

asteroids and comets. These missions involved n only taking observations from a distance whi orbiting the planet, but also, in some cases li Mars and Venus, landing probes on the surfac and, in the case of Jupiter, releasing probes in the atmosphere.

By the 1970s, the American obsession wi lunar landing was over and the focus shifted to t outer planets and an orbiting space station. T Soviets launched their first space station, Salyu I, in 1971 while the Americans constructed t Skylab in 1973. In 1975, there was a successf docking of the Apollo and the Soyuz spacecraf signaling a collaboration between the two riv space powers.

The Soviets launched several Salyut spa stations to test the technology and, finally 1986, launched the Mir, a modular space static which could be built in stages. It was a remarkab successful endeavor—despite having a predicte life of only five years, it stayed in orbit for over 1 years. In 1998, a truly international effort, th International Space Station (ISS) was launche and is still in orbit. This 16-country collaborati enterprise is among the most ambitiou technological projects in history.

After the Apollo mission, the next ambitiou project taken up by NASA was to develop reusable spacecraft. This was supposed to ser as an affordable transport vehicle to supply th space stations being planned as well as to launc satellites. The Space Shuttle took almost nir years to design and test. Finally, in 1981, th Space Shuttle carried two astronauts into a nea earth orbit. Over the next two decades, the Shutt would prove to be the mainstay of the America space program, though it was not as inexpensiv as the projections.

THE SPACE RACE

The space race between the US and the Soviet Union was responsible for many achievements in space exploration. It began in 1957 with the launch of Sputnik I by the Soviets and intensified when Yuri Gagarin orbited the earth in 1961 in Vostok 1, becoming the first human in space. The US reacted with the announcement by President Kennedy in 1961 that the US would land a man on the moon before the end of the decade. This set into motion a huge research and development project—Project Apollo was put in place with a massive increase in the funding available to NASA for the purpose.

The basic design for the proposed lunar landing was decided in the early 1960s. This included using the powerful Saturn V rocket to deliver the Apollo spacecraft into outer space. The spacecraft itself was designed to have two main sections: the Command/Service Module for take-off and landing back on earth, and the Lunar Module for landing on the moon and then taking off again to reunite with the Command Module.

The next few years were used to design and manufacture the components needed for this ambitious project. More than 30,000 employees at NASA and more than 300,000 workers at the various sub-contractors were involved in the lunar landing program. After some initial setbacks, in 1968, Apollo 8 mission was launched and the spacecraft had a successful run. It circled the moon and came back with its three astronauts carrying back photographs of the surface as well as of earthrise over the lunar surface. The next two Apollo missions, Apollo 9 and Apollo 10, were used to carry out tests of all the critical components in flight and during take-off and landing. Apollo 10 was in fact a mini-rehearsal for the actual landing on the moon and it came very close to the lunar surface.

Finally, on July 16, 1969, Apollo 11 carrying three astronauts, Neil Armstrong, Edwin Aldrin and Michael Collins, took off from Cape Canaveral. Four days later, on July 20, the lunar module carrying Armstrong and Aldrin landed at the Sea of Tranquility on the lunar surface. Within a few hours, Armstrong left the Lunar Module to step on the moon and became the first human to have set foot on another body in the universe.

The astronauts picked up rock samples and returned to the Command Module after about 12 hours. Four days later, the Command Module splashed safely into the Pacific, thus bringing to a happy end the first manned lunar odyssey. Television and radio had made possible for hundreds of millions of people around the globe to witness this magnificent event. The footprint left by Armstrong on the lunar soil became a symbol of technological prowess of mankind in its quest for exploration.

Apollo 11 was followed by six more missions to the moon. Apollo 13 mission was a disaster when the oxygen tank exploded, though the crew returned safely. The last four missions also carried with them lunar rovers which allowed the astronauts to drive around the lunar surface. Photographs and samples of lunar surface were collected for investigation by scientists back on earth. These investigations would enable the scientists to get a better understanding of the birth and evolution of the moon. The Apollo 17 mission was conducted in December 1972 and this brought an end to the exploration of the moon.

Neil Armstrong

Like Yuri Gagarin, Neil Armstrong also captured the imagination of a whole generation. Armstrong's first space flight was aboard Gemini 8 in 1966, when he performed the first space docking between two vehicles. In 1969, he commanded the Apollo 11 mission to the moon and became the first human to land on the moon. Armstrong and several others after him studied the lunar surface, left instruments there to gather data and also collected lunar soil and rock samples for analysis.

After the success of the Apollo mission, both the US and the Soviet Union shifted their focus from sending man to the moon to developing a much more efficient strategy for space exploration. The Soviets had already began developing orbiting space platforms where cosmonauts could stay for extended periods of time and perform experiments. Orbiting space platforms were also an essential part of the strategy for setting up manned exploration of distant planets. This was because these platforms would be used as a first stop for venturing further out into space.

The first Soviet space station, Salyut I, was launched in 1971 and this was followed by the launch of the American Skylab in 1973. The Soviets launched several other space stations as part of the Salyut series and, finally in 1986, launched the Mir space station. Over the next few years, other modules were added to the core module of the Mir and several cosmonauts from a number of countries visited the space station. It was a hugely successful venture and orbited for more than 14 years before finally coming down to earth in a controlled burn off.

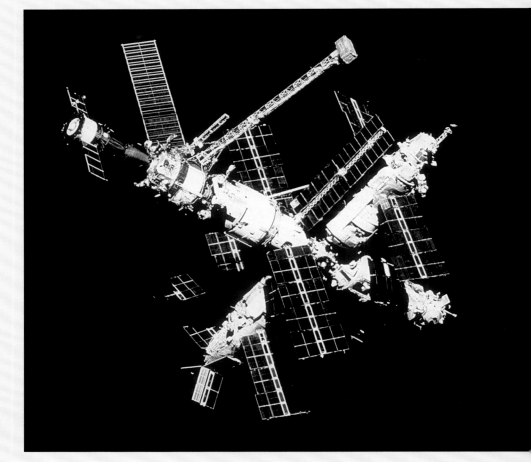

Mir space station

In 1971, the Soviet Union launched the first space station Salyut for carrying out research programs in space. Salyut was followed by a more advanced space station Mir, which was launched in 1986. Mir continued in operation till parts of it were incorporated into the International Space Station (ISS). The ISS is a modular space station which is being constructed since 1990 with international collaboration.

In 1984, the US launched its own space station program but this time as an international collaboration. The successful development of a reusable launch vehicle, the space shuttle, had already marked the first step of building very complex platforms in space for extended use by humans. This became necessary because the task of placing the large amounts of equipment in orbit ordinarily required huge amount of resources if the spacecraft were not reusable. With the shuttle coming into operation, this problem was solved since men and material could now be carried to the space station quickly and with a significant reduction in cost. Six space shuttles have been built so far, and it is proposed that the space shuttle would be replaced by a new vehicle Orion after 2010.

With the collapse of the Soviet Union in 1989, the Soviet space program suffered for some time. In 1991, the US invited the Russians to join in the International Space Station (ISS) which was to form the mainstay of the space program for the next century. The first couple of modules of the ISS were launched in 1998 and there are plans to continue launching modules for many years to build up the space station.

The Apollo mission to the moon was able to demonstrate both the capabilities and limitations of manned exploration of outer space. Landing a man on the moon, though a hugely ambitious task, was nothing compared to the challenges faced by sending manned spacecraft to other planets in the solar system. The technological issues are so enormous that there are at present no plans of ever attempting such an audacious venture.

Manned exploration has always had both cost and benefits. Sending humans on a space mission is many times more expensive than sending unmanned spacecrafts. To ensure that the astronauts are safe during the mission is a major task, though even the task of sustaining human beings in space for extended periods, which would be required for interplanetary trips, is not easy. The risks involved in manned missions were clearly shown by the many accidents which have taken place since Yuri Gagarin orbited the earth. The most recent one in which seven astronauts were killed, when space shuttle Columbia exploded during re-entry into the earth's atmosphere on February 1, 2003, is possibly the most dramatic and tragic of them. However, it is also argued that the capabilities of human beings in carrying out experiments and repair of equipment in space is unmatched by instrument howsoever sophisticated.

With increasing pressure on the resource of earth, it might be possible that the future of mankind lies in establishing habitation on other planets and astronomical bodies. Mining the moon or establishing colonies in space may have seemed fantastic at the time of Jules Verne, but not such an impossible dream for humans now.

Transportation

In 1955, the English engineer Christopher Cockerel invented the hovercraft, a vessel which moves over land or water without touching the surface. The first hovercraft crossing of the English Channel (by R-N1) took place in 1959 and regular services across the Channel began soon afterward.

In aviation, the introduction of the jetliners into commercial service in the 1950s was a big step in air travel as this increased the capacity as well as the range of the aircraft. Supersonic aircraft were being used for reconnaissance and other military purposes. In 1975, the first commercial flight of the supersonic Tu-144 cargo plane, and then the following year of the Concorde, brought this technology into the civilian domain.

Automobiles also underwent significant design advances in this period. While the 1950s and 1960s were the era of the dominance of high-powered engines which consumed enormous amounts of fuel, the twin pressures of environmental movements and the oil shock of the 1970s led to the development of more efficient engines. Electronics, especially microprocessors, controlled various functions of the automobile engine. The revolution lies not only in engine technology but also in materials used for building the engine and the body—lightweight materials such as aluminum, plastic and carbon fibers were increasingly used to reduce the weight of the automobile.

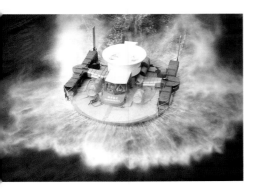

Molecular Biology

The developments in molecular biology in the second half of the 20th century had a far-reaching impact on humans. It all began with the discovery of the structure of the genetic material DNA in 1953 by two Cambridge scientists, James Watson and Francis Crick. Although a lot of work in trying to understand genetics at the cell level had taken place in the previous years, it was the brilliant insight of Watson and Crick that put the pieces of the puzzle together.

Even though the basic structure of the genetic material was enunciated in 1953, the actual details of how the cell functions, especially the formation of proteins, was not very clear. In 1957, Crick proposed the Central Dogma of Biology which

helped explain the relationship between the key elements, namely DNA, RNA and proteins. The actual solving of the genetic code came only in the 1960s with the work of Marshall Nirenberg and Har Gobind Khorana, who showed how the sequence of nucleotides on the DNA produced the different amino acids.

Even though the basic mechanism of genetic transmission was understood by the end of

1960s, the actual revolution came later with the invention of several techniques which made possible the manipulation of genetic material. In 1970, American microbiologists Daniel Nathans and Hamilton O. Smith discovered enzymes which could snip the DNA molecule at predetermined sites. These restriction enzymes proved to be among the most important elements for the future of genetics and molecular biology. Using restriction enzymes, Paul Berg in 1972 was able to use genetic material (DNA pieces) from different sources and splice them together to create a recombinant DNA molecule. Several techniques were developed to speed up the process of determination of the nucleotide sequences on a DNA strand during the 1970s. The biotechnology industry, which was mostly focused on recombinant DNA technology, started to manufacture useful genetically engineered chemicals.

The next big breakthrough came in 1983 when the American biochemist Kary Mullis invented the Polymerase Chain Reaction (PCR), a technique which allowed scientists to very rapidly identify and replicate a specific DNA sequence. It was this technology which laid the groundwork for the huge explosion of the biotechnology industry.

Among the most visible results of biotechnology is cloning. Frogs had been cloned in the 1950s and mice in the 1980s, but notably all by use of embryonic cells. The first big success of

Dolly the cloned sheep

The first animal to be cloned from an adult, non-reproductive cell was a sheep, which was called Dolly. The successful cloning of a sheep from the mammary cells demonstrated the potential of cloning techniques and also opened up debates on the ethical issues regarding cloning. Created at the Roslin Institute in Scotland, Dolly lived for almost seven years before being put to sleep in 2003.

SR-N1 hovercraft

Although several people had tried using an air cushion to reduce the friction in boats, it was in 1952 that the British inventor Christopher Cockerel came up with the workable design of the hovercraft. The first prototype SR-N1 (Saunders Roe Nautical 1) was made in 1959. It was a huge success and even crossed the English Channel. Hovercrafts came into regular service some years later. Here SRN-1 is seen on the Thames at Westminster.

cloning from an adult animal was in 1996 when researchers in Roslin Institute in Scotland were able to successfully clone a lamb called Dolly from adult sheep cells. Although ethical issues abound in cloning, the technique could prove to be of immense use in animal breeding, medicine and the pharmaceutical industry.

Agriculture

Though molecular biology was expanding horizons of human knowledge, it was only after the 1980s that its potential in agriculture, pharmaceuticals and medicine could be realized. In the meantime, largely due to the efforts of the Ford and Rockefeller Foundations, a unique experiment in Mexico had been successful in producing high-yielding, disease resistant varieties of wheat and maize. The man behind this was the American agricultural scientist Norman Borlaug. The high-yielding varieties (HYVs) soon spread to many parts of the world, notably the Indian subcontinent where agricultural production increased dramatically, reducing the risk of famines. This enormous improvement in food grain production, achieved by new crop cultivars, fertilizers, pesticides and mechanization in agriculture, caused the Green Revolution of the 1960s and 1970s.

After the biotechnological revolution, agricultural scientists realized the huge potential of genetically modifying plants to make them disease or pest/herbicide resistant. These genetically modified organisms (GMOs) have been made by inserting genes which make corn, cotton and soya bean pest resistant. There has been a huge debate regarding the health and environmental safety

of these transgenic crops but, at the same time there is an increasing demand for these varietie especially in the developing countries.

Medicine

The tremendous developments in medicine especially in orthopedics and surgery, durin the world war were followed up by even mor radical innovations in the field of medicine pharmaceuticals, diagnostics and surger All through this period, new antibiotics wer synthesized, enabling the fight against a variet of microbes. Vaccines against bacterial disease like tetanus and typhoid were already available In this period, these were made more effectiv and safer. In addition, vaccines against variou viral diseases like polio, influenza and measle were developed. In 1954, Jonas Salk made th first polio vaccine, and in 1960, oral polio vaccir was developed by Albert Sabin. Vaccinatio against various infectious diseases played a majc role in decreasing the morbidity and mortality the West.

Although most of the infectious diseases ha been tackled by the end of the 20th century there remained a major challenge for medicine namely the treatment of cancer. Malignancy ha been treated with radiation and X-rays for a lon time. After the development of nuclear reactor various radioisotopes were made available t ensure more effective radiation therapy. I addition, chemotherapy with very potent an effective chemicals has proved to be of immens use in the handling of malignancy of various kinds Though there is no effective cure, except for thos

CT scan and MRI

Huge advances were made in medical diagnostics in the 20th century with the advent of techniques like Computerized Axial Tomography or CAT scan (invented by Godfrey Hounsfield and Allan Cormack in 1971), Magnetic Resonance Imagining or MRI (patented by Raymond Damadian in 1974) and Positron Emission Tomography or PET scan (invented by Edward Hoffman, Michael Ter-Pogossian and Michael Phelps in 1973). These techniques allowed the medical practitioners and researchers a much more detailed view of the human body than was possible with X-rays. The development of these techniques was made possible because of the tremendous progress in computer and electronic technology. Here a CT scan and MRI scan of the human heart, brain and the spine are shown.

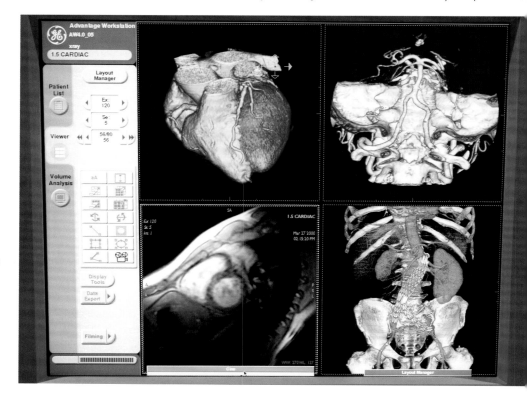

ases where surgical intervention is possible, the management of the disease has reached such a stage that life expectancy has increased hugely in cancer patients.

In the field of surgery, two developments stand out from this period. One was an increased use of new materials like plastics in a range of applications—from improved sutures to replacement lenses in cataract patients, to heart valves and orthopedic supports. Secondly, the invention of microscopic surgery proved to be of great use in removing post-operative complications among patients. In the case of heart surgery, the development of the heart-lung machine allowed surgeons to operate safely on patients. The use of stents and balloons made the surgical cure of heart diseases fairly commonplace. Artificial pacemakers to regulate the rhythm of the heart were also introduced in this period. Though the idea of a pacemaker had been around for several years, it was only after the development of the transistor that portable pacemakers became a reality in the late 1950s.

Another area where significant developments took place was the field of diagnostics. The 1970s saw the invention of Computerized Axial Tomography (CAT), a technique which allowed physicians to get a three-dimensional image of the internal organs. The technique involved taking two-dimensional X-ray images from various angles and then reconstructing the three-dimensional image using computers. It was invented independently by Godfrey N. Hounsfield and Allan M. Cormack in the US. This was an extremely important development, especially concerning the imaging of the brain.

In 1970s, another technique, called Magnetic Resonance Imaging (MRI), was introduced by Raymond Damadian. This diagnostic procedure employed nuclear magnetic resonance to obtain sectional images of the internal structure of the body. Used in chemistry to study molecules that make up a substance, the principle was applied in medicine to produce three-dimensional images of the soft tissues of the body, especially of the nervous system from the brain to the spine. Compared to CAT, MRI produces a far more detailed image and has no side effects.

The improvements in pharmaceuticals, diagnostics, surgical techniques and public health massively decreased the mortality and morbidity rates in the developed as well as developing countries. The wide-ranging immunization drives against some of the most deadly diseases led to their near eradication. However, several challenges like the HIV, malaria and drug-resistant tuberculosis still remain to be tackled. In addition, at least in the West, lifestyle problems like heart diseases have emerged as the biggest causes of mortality.

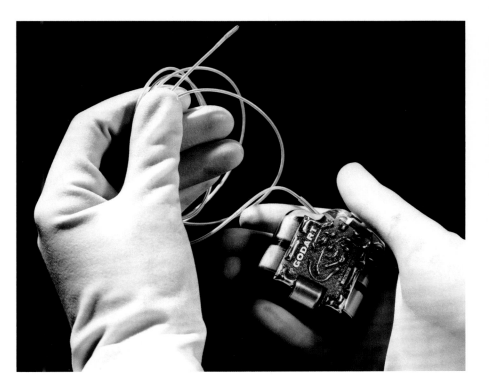

Plastics and New Materials

Although plastics had been invented and were in use before the world war, they were used in a variety of fields during this period. Different qualities of plastics were invented, including those that could be used safely in medicine as implants. Apart from plastics, this period saw the invention of a revolutionary new material—carbon fullerenes. In 1985, two groups independently discovered this new form of carbon (apart from coal, graphite and diamond, the known allotropes of carbon) in their experiments with molecular beams. This class of compounds had a very unusual structure and resembled a geodesic dome that had been popularized by the architect Richard Buckminster Fuller, hence the name fullerenes.

Subsequent experiments revealed several different kinds of fullerenes, namely buckminsterfullerene and carbon nanotubes. These substances have very unusual properties which have already found many applications in nanotechnology, electronics, cosmetics, optics and other fields of materials science. The discovery opened up a whole new field of analyzing and manipulating matter at very small, nanometer (a billionth of a meter) scales.

Semiconductor technology also witnessed several innovations in the post-war years. Better, more efficient semiconductors were developed to be used in electronic devices. They also found use in photo or solar cells. The launch of space probes and satellites meant that a reliable source of power was needed to sustain the electronic circuits within. Solar panels, made of photovoltaic materials, were (and continue to be) used for this purpose. Though photovoltaics have been known for over a

Pacemaker

In 1957, the first external wearable pacemaker was made by Earl Bakken, using the recently developed transistor technology. The pacemaker was a very useful device for patients with heart ailments. It could regulate the beating of the heart by using electrical impulses. Within a few years, implantable pacemakers with a much longer life were developed. The photograph shows an early heart pacemaker made of epoxy resin and powered by mercury batteries with a life of three years.

century, it was in the years following the launch of the Sputnik that more efficient and practical solar cells were developed.

Nuclear Power

The big innovation in the field of power was the introduction of nuclear power. After the

Three Mile Island nuclear plant

After the successful demonstration of a controlled nuclear chain reaction during the war years, reactor research was mostly for military purposes. In 1951, electricity was produced for the first time using nuclear reaction. The nuclear industry saw a rapid growth during the 1960s and the 1970s with some countries like France using nuclear power on a massive scale. However, nuclear leaks at the Three Mile Island nuclear plant near Harrisburg, Pennsylvania in 1979 and at Chernobyl in 1986 led to a slowdown in the nuclear industry.

The first American ICBM

The missile and rocket technology developed in Nazi Germany was used by both the Americans and the Soviets to develop long range Intercontinental Ballistic Missiles (ICBMs) in the 1950s. Over the years, technology advanced rapidly, leading to more accurate missiles with capabilities of attacking multiple targets with the same missile. Here a group of children watch the first American ICBM—an 80-ft (24-m) missile—rising into launch position in 1959 at the Aquarium Compound, Coney Island, Brooklyn, New York City.

development of the atomic bomb, it was realized that the amount of energy generated in a controlled atomic fission could be used to generate steam and thus run turbines to produce electricity. In 1953, the Atoms for Peace program was launched in the US which laid the framework for peaceful use of nuclear power.

Although the basic principles of harnessing the tremendous energy of nuclear fission were well known, the actual design of reactors to take care of the safety requirements was quite complicated. In the late 1950s and 1960s, many designs were tried and several commercial nuclear power plants were commissioned. These included pressurized water reactors, light water reactors and gas cooled reactors. The nuclear power industry got a further fillip with the oil shock in the early 1970s when there was a threat to the energy security in the West. However, a series of accidents—notably the Three Mile Island accident in the US in 1979 and the Chernobyl disaster in the Soviet Union in 1986—made the world aware of the dangers of nuclear energy. Even after those experiences, there's been little development in this matter.

Although fusion reactor technology was well established by the end of the 20th century, there was no real breakthrough in making nuclear fusion practical and useful. Fusion, which works on the principle of fusing of light nuclei (as opposed to

fission where energy is released when a heavy nucleus splits up into lighter ones) to generate energy, has the potential of being an essential limitless, clean source of power. However, the technical issues involved in making a practical fusion reactor have proved to be formidable, though much progress has been made in understanding what might be needed to bring this to fruition.

Military Hardware

The world war had demonstrated the devastating effects of technology—the V-2 rockets, the improved aircraft capable of carrying large payloads, more destructive bombs and artillery and finally the atomic bomb. The arms race between the US and the Soviet Union, which was a constant feature of the post-war period till the collapse of the Berlin Wall, followed as a simple corollary to the cold war.

The atomic bomb had already been tested during the last stages of the war with Japan. The hydrogen bomb, a much more lethal weapon, was first developed and tested by the US in 1952, and soon the Soviet Union responded by testing its own fusion weapon. Once the basic principle of these catastrophic weapons had been tested, there was a race for designing and developing bigger and more lethal bombs of the same kind. These weapons were deployed on land, in air and on sea to ensure that there were enough of them left in case the enemy attacked first. This strategy of deterrence led to a huge increase in the quantity and quality of weaponry and put the world on alert. Now there was a need for a reliable and accurate delivery system.

Right after the world war, unguided missiles capable of attacking enemy positions had been developed. These were deployed in western and eastern Europe in the 1950s. However, the developments in electronics and navigation systems during and after the war were soon incorporated into missiles. Over this period, a variety of tactical guided missiles were developed and deployed both by the US and the Soviet Union. Most of these carried thermonuclear weapons which were being developed at a huge rate.

The qualitative change came with the development of strategic missiles which were capable of attacking long distance and could be used in lieu of manned aircraft. Another advantage was that they could be deployed all over so as to resist enemy attacks. There were two kinds of strategic missiles—ballistic missiles which were rocket-powered and hence flew at a very high altitude to avoid detection, and low altitude flying cruise missiles using jet engines.

The technology used by the Germans during the war for making the dreaded V-2 rockets was used by the Americans after the war. Throughout the 1950s, there were many attempts to make a long-range missile capable of carrying heavy nuclear warheads. Finally in 1959, the first operational ICBMs (Intercontinental Ballistic Missiles), Atlas and Titan, were deployed in the US. These were liquid-fueled and had limited capabilities in terms of payload and range. Urged by the need to surpass the Soviet missile technology, a sample of which was seen in the launch of Sputnik in 1957, the US sought to develop a solid-fueled ICBM. The efforts led to the development of Minuteman, which took the US ahead of the Soviet Union in the missile race. The transition from liquid to solid fuels allowed powerful engines, greater ranges and increased "safety" in missiles. Moreover, these missiles were more cost-effective, smaller and better suited to mass production.

The 1970s saw another major development in ballistic missiles when multiple warhead systems were introduced into the arsenals on both sides. These were very deadly weapons carrying multiple warheads which could attack multiple targets, thereby increasing the destructive power of the missile.

Finally, during the 1980s, the US launched the Strategic Defense Initiative or SDI, which attempted to use various kinds of weapons to destroy in the air the nuclear missiles from the enemy. These weapons included laser weapons and collision weapons.

The cold war ended in 1991 with the collapse of the Soviet Union and thus these weapons were never deployed, though their development probably went on.

Communications

The invention of the transistor in 1947 at the Bell Labs ushered in the semiconductor and electronic revolution. During the next few years, progress was made in using semiconductors to create capacitors and resistors, and putting these together with transistors and diodes on the same silicon piece. This way, complex devices could be miniaturized to a large extent, though the problem of interconnecting still remained.

A big breakthrough came in 1958 when Jack Kilby and Robert Noyce independently showed that it was possible to lay extremely thin metallic channels directly on the silicon piece on which the other devices had been fabricated. This way, the whole circuit, including the interconnects, could be integrated into a really miniature device.

Integrated Circuits, as these came to be known, were the key components of the microelectronic revolution which took shape during this period.

Sound recording and reproduction had undergone several improvements with better and higher quality phonograph records during the previous decades. However, a new medium was introduced in the early 1950s, which essentially replaced phonographs. This was the magnetic tape, which was initially invented by German engineers during the world war for sound recording. The original design was taken by the British and Americans and improved to make a high-quality sound recording and reproduction system. The popularity of the new medium was tremendous, and within a few years, tape recorders proliferated into a large number of households in the West.

Similarly, though there were many innovations in the field of motion pictures in the previous decades (such as the introduction of sound and color),

A videotape recoder

In 1951, Charles Ginsburg and his team invented the videotape recorder, which captured the television images and converted them into electrical impulses to be saved onto magnetic tapes. With the sale of the first video recorder in 1956, it became possible to record video and audio programs at a fraction of the cost of a film. However, in the initial years, the technology was too expensive for the home market and was confined to television networks and businesses. It was only in the late 1970s that an affordable videocassette recorder for the home market was developed. Video recorders became hugely popular during the 1980s and defended their position till they were replaced by compact disc technology.

television was still transmitting live programs since there was no way to record shows and then run them later. In 1956, Charles Ginsburg and Ray Dolby invented a practical video recorder which used magnetic tapes. This was a major innovation for the television industry which now could, like the radio, record and air programs at will. After the initial introduction of video recording

Stack of CDs/DVDs

Though the compact disc was invented by James Russell in 1965, it was mass produced only in the 1980s. The technology was an instant hit. The use of lasers to digitally record and replay music brought about a quantum leap in the quality of sound reproduction. The subsequent introduction of Digital Video Discs (DVDs) meant that video recorders could be replaced by much higher fidelity recordings.

for professional use, there was the development of a portable and more economical kind of video recorder to be used for home entertainment. Video recorders became hugely popular in the 1970s and the 1980s, and pre-recorded video tapes of movies and television shows became ubiquitous.

Both the audio and the video tape technology exploited the magnetic properties of certain kinds of materials to record audio and video data. However, in 1980, Phillips Corporation and Sony Corporation introduced a new way of recording data in digital form. The compact disc or CD used laser light to read and write on a plastic media. Though the CD was invented in 1957 by James Russell, it became popular after it was mass produced in the 1980s. CDs were undoubtedly the most important development in sound and data storage since the phonograph.

In 1982, CDs were adopted for digital audio recording and were hugely successful since they could store more data and with a much better quality. Moreover, the quality did not deteriorate over time. Over the next few years, video CDs, which used the same technology to record video and sound, were also introduced, but these were not as successful as audio CDs because the data that needed to be stored for high quality video was very large.

In 1995, once again Phillips and Sony introduced a digital video disc or DVD, which could store much more data than a laser disc or an audio CD. For the first time, it became possible to digitally record and reproduce video and audio in very high quality. The key to this technology was the use of high-powered lasers which allowed a much smaller pit size to record the data.

Photography also saw various innovations and improvements during this period with more vibrant colors and better fidelity in the images. However, all these innovations did not alter the basic technology of taking pictures and recording images—a film was used to record images which then was developed to make positives. A revolution in photography came with the introduction of digital cameras when the basic format of pictures got changed.

The first filmless digital camera was made at Eastman Kodak by Steven Sasson in 1975. However, this camera was too bulky and slow, and never got into production. In 1990, the first commercially available true digital camera was introduced. This used a charge coupled device (CCD) to record the image which was then stored as a digital file to be read on a computer. Soon, different kinds of digital cameras were introduced, including single lens reflex or SLRs.

Much like photography, the basic telephone also underwent radical changes in the second half of the 20th century. By the 1950s, a fairly widespread telecommunication network had been established in most developed countries. There were of course various improvements and innovations in the switching and transmission equipments as well as in the telephone receivers. With the launch of the space satellites, it was quickly realized that one of the most important applications of satellites could be for communication technology.

In 1962, the first communication satellite, Telstar 1, was launched in a medium earth orbit. It was the first satellite to transmit television broadcasts and telephone conversations across the Atlantic. Syncom 2, launched in 1963, was the first geostationary communications satellite Placed in the geostationary orbit—at an altitude of 22,000 miles (35,400 km)—and orbiting the earth at the same speed as the rotation of the earth these satellites appear to be stationary above the same location on earth. This technology allowed non-stop communication between the ground stations and the satellite. In the next few decades, the number of geosynchronous communication satellites increased to 300, with each of them equipped to transmit enormous data.

The other great innovation in transmission was the introduction of fiber optic cables. The fact that light can be guided through an appropriately

constructed glass fiber had been known since the 19th century but the technology to make it viable was unavailable. In the 1950s, researchers were trying to make a flexible gastroscope for diagnostic examination of the stomach. In the process, during the late 1950s, they made the first glass-clad fibers.

The invention of the laser in 1957 led researchers to think about the use of fibers for communications. In the 1960s, several innovations were made in the field of glass technology to get low attenuation rates in the fiber. These were coupled with innovations in the development of optical amplifiers in the 1970s which allowed signals to be transmitted over long distances.

By the 1990s, the need to transmit large volumes of data provided further impetus for the industry and a lot of optical fiber was laid for communications. Fiber optic has the advantage of being able to carry enormously large amounts of data (whether voice, data or video) with little loss over large distances.

The combination of radio and telephone technology had been tried for specialized applications during the earlier years, especially during the world war. In fact, the first mobile telephone service was launched in 1946 in the US. This was a very rudimentary service which allowed radio transmitter and receivers to communicate with fixed telephones. This was improved upon in 1964 with the introduction of the Improved Mobile Telephone Service. Though a vast improvement over the earlier system, this still had severe limitations in the amount of traffic it could handle.

The first mobile handset was invented in 1973 by Motorola, though the handset was very bulky

and impractical. In 1984, Bell Labs developed the call hand-off feature which allowed true mobility and very efficient use of radio spectrum for telephony. In 1991, the first second generation, true mobile network was launched in Finland. The stupendous improvements in the switching technology and in mobile handset technology, mostly due to the increased use of digital technology, followed rapidly. By the end of the century, mobile telephone penetration was almost reaching the fixed telephone penetration in several European countries.

The possibility of sending moving pictures together with voice over the telephone had been investigated since the late 1920s, though not much progress was made except in experimental set-ups. In 1956, Bell Labs developed a videophone which could be used over the telephone network. This was further improved and a second-generation picture phone was introduced in 1971. With the development of image compression techniques as well as faster and cheaper computing power, digital videophones were introduced by AT&T in 1992. These allowed full motion transmission and thus were a significant improvement over the previous technology.

Computers

The field of electronics and computers saw an exponential growth during this period. The key component, as we have seen, was the invention of the integrated circuit which allowed miniaturization on a huge scale of components. After the successful deployment and use of the ENIAC in the 1940s, a lot of work went into making computers more powerful and useful for a variety of applications.

Mobile telephony

The concept of cellular phones was first introduced by Bell Labs in 1947 with the police car technology. In 1973, Motorola, using the same technology, developed the first portable phone which could be used outside a vehicle. Although mobile phones became commercially available in the early 1990s, the explosive growth in penetration came in the late 1990s and the early part of the 21st century. The revolution brought about by this device has been unmatched in the history of telecommunications and there are now many more mobile subscribers in the world than fixed-line ones. The technology has advanced rapidly and the networks and devices now offer video, Internet and other features.

Early computers

The early mainframe computers were mammoth devices, weighing many tonnes and taking up a huge space. UNIVAC I (Universal Automatic Computer I) was made in 1951 and designed for administrative and business use rather than scientific use. The photograph shows a UNIVAC I being tested by the US Air Force. The machine took up more than 300 sq ft (28 sq m) of space and ran at about 2 MHz.

Punched cards

The earliest punch cards were used in textile looms in the 19th century and were later used in calculating machines in the late 19th century. In the early years of digital computers, punched cards were used to input and output data and programs. IBM developed the technology and IBM machines to record data on punched cards became ubiquitous. The photograph shows an operator working on an IBM Translator.

Intel Pentium processing chip

In 1993, Intel launched the Pentium or P5 series of microprocessors to succeed the 486 processors which were being used. Based on a radically new architecture, the Pentium allowed much faster data processing and was the dominant architecture for almost a decade in most personal computers.

The seminal theoretical work on this had already been done by the Hungarian–American mathematician John von Neumann who had been associated with the ENIAC as well as the Manhattan Project to build the first atomic weapon. In 1946, von Neumann, together with his colleagues, wrote a seminal paper which laid out the foundations of computer science. The key insight of von Neumann was to store the input data and instructions for the computer together to enable faster access.

The first computer with stored program was made in Britain in 1949 but it was not very successful. The first truly successful stored program machine was built in 1951 for the US census bureau. The UNIVAC I (Universal Automatic Computer I) was a commercial data processing computer with a keyboard. With the input and output data stored on magnetic tapes, the machine was the fastest business machine in its time. Although much smaller than the ENIAC, it still occupied almost 874 cu ft (24 cu m) of space.

The next few years saw the development of several machines by manufacturers like Remington Rand, Burroughs and, most importantly, IBM. All these machines were based on the stored program technology and each one showed an incremental advance over the others. However, the use of computers was still very restricted since operating a computer was a highly specialized job and the machines themselves were very expensive. Using the machines was also very time-consuming since they could only process one program at a time; this was a crucial bottleneck.

Another major obstacle for the development and use of computers was the unavailability of programming languages. All the instructions for the computers had to be written in what was called a machine language which was extremely tedious for general use. The need for a translator, a system which could convert ordinary mathematical operations into machine language understandable by the computer, had been felt for a long time but there was no real progress in this direction.

In 1952, Alick Glennie, a student at the University of Manchester, created a prototype of such a program which he called Autocode. In 1954, engineers at IBM designed what was to be the first and extremely popular high level computer language, FORTRAN. By 1957, IBM had released the compiler (translation program) for FORTRAN. For the first time, ordinary non-specialist users could write programs to be run on the computers. Subsequently, many other languages were developed, but FORTRAN continued to be popular, especially among scientists and engineers.

IBM was also responsible for several key inventions which made computers more accessible and powerful. The magnetic disk and the chain printer were two such inventions. IBM had also started using the newly developed integrated circuit technology to make the machines more versatile. An example of IBM's innovation during this period was the bestselling IBM 360 series which dominated the computer market throughout the late 1960s.

Another development was the introduction of time-sharing, i.e. the use of the computer resources by more than one program. Although a lot of theoretical and practical research work had been done on this subject, it was only in the mid 1960s that the concept really became practical. A key invention in this sector was the development of a new operating system, UNIX, by two Bell Lab researchers, Dennis Ritchie and Ken Thompson. UNIX was a very powerful operating system and allowed a lot of flexibility to the programmer. It was distributed freely to universities and this added to its popularity. Various versions of the UNIX operating system evolved with different features and it continues to be the powerhouse of servers.

In 1965, a new kind of machine, called minicomputers, was developed and further expanded the use of computers. These were typically stripped down versions of the bigger machines (now called mainframes) but with the same basic functionalities. Minicomputers were more affordable than mainframes and soon most universities and companies adopted them and continued to use them till the mid-1970s.

The microprocessor revolution changed the rules of the game. Integrated circuits were already being made during the 1960s and were becoming more and more sophisticated with an increasing density of components packed into single devices. In 1971, Intel, a young company in the area around San Francisco (later coming to be known as the Silicon Valley), developed an integrated circuit which could process instructions. The 4004, as the device was known, was used in calculators by a Japanese company and sold well. But the first real, usable microprocessor was the 8008 which was a 8-bit processor unlike the 4-bit 4004. Introduced in 1972, the 8008 was improved to the 8080 in 1974. However, no one really knew what could be done with these revolutionary devices.

The first microcomputer Micral was developed in France in 1973–74 using the 8080 microprocessor but this was not very well known outside France. In 1975, a company called MITS produced a hobby-kit to make a computer called Altair. This was enormously popular with hobbyists and, in fact, most of the innovators in the computing field were avid Altair hobbyists.

In the field of software, the first spreadsheet program called VisiCalc was developed by Dan Bricklin and Bob Frankston, and Gary Kildal released the first operating system for microcomputers CP/M (Control Program for Microcomputers) in 1974. Two Altair hobbyists—Bill Gates and Paul Allen—formed Microsoft Corporation to develop software for microcomputers. Two other hobbyists, Steve Jobs and Stephen Wozniak, also joined hands to make a computer in Jobs' garage. The result of their effort was the first Apple machine which they sold in very small quantities. However, they soon formed a company and came up with huge successes like Apple III, and finally in 1982, introduced the first computer which used Graphical User Interface (GUI). This computer, called Lisa, was revolutionary but not very successful. GUI technology had already been invented—like a host of other innovations (the mouse, laser printer, Ethernet and so on)—at the Xerox research center at Palo Alto (PARC), but Xerox had failed to commercialize these technologies.

In 1984, Apple introduced a scaled down version of the Lisa machine, called the Macintosh. This was the first truly successful personal computer because of its design and ease of use. Earlier, IBM had entered the microcomputer business by introducing its personal computer. It used a 16-bit 8088 Intel processor and was capable of doing various things like word processing and spreadsheets. Using an operating system sold by Microsoft Corporation called Disk Operating System or DOS, the machine was more powerful than anything available in the market and sold more than half a million units in the first two years.

The next few years saw many developments in personal computers, both in the hardware and the software. Thanks to Intel's improved processors,

Steve Jobs and Apple Computers

Steve Jobs, a high school graduate, together with Steve Wozniak, designed the first Apple computer Lisa in 1976. Jobs and Wozniak went on to develop the Macintosh computer, the first computer to use the Graphical User Interface which made working on it much simpler than the IBM Personal Computer. Jobs left Apple for a few years, and on coming back, rejuvenated the company with new products like the iPod and iTunes.

more performance could be packed into the personal computer. Higher density of magnetic hard disks meant more data could be stored, while the introduction of devices like the floppy disk drives meant easier portability of data.

Portable computers or hand held devices also became popular during the 1980s and were a big hit during the 1990s. In their earlier avatars, these devices tried to do what a personal computer would do, but only tried to shrink the size so as to make the device portable. But this was not much of a success.

In 1996, Palm Computing Inc introduced the Palm Pilot, a unique hand-held device which did not replace the personal computer but instead allowed the user to have data readily available. It had a stylus and could recognize handwriting by using special software. It was light and small in size and could easily be connected to personal computers. Because of its popularity, by the end of the century, several computer manufacturers were trying to introduce their own versions of mobile devices.

World Wide Web

The world wide web was developed at CERN in Geneva in 1988 to facilitate data exchange between multi-national collaborators on the particle physics accelerators. With the introduction of the first browser Mosaic in 1993, the Web became accessible to everyone. The World Wide Web, together with the Internet, revolutionized the way information is shared and accessed by a large fraction of humanity, especially in the developed countries. This globe of web pages represents globalization of information made possible by the Web.

Microwave oven

Cooking with microwaves was discovered during World War II when some scientists were doing research on microwave radars. Although the technology was patented in 1945, it never really became popular. It was only in the late 1970s that the technology improved enough to make microwave ovens affordable and usable in ordinary homes. By the end of the century, microwave ovens were to be found in almost all homes in most developed countries.

Internet

By the end of the 1980s, personal computers had become quite popular in the West and were being used extensively in business and academia, too. They were getting more powerful with more and more features added every year. The peripheral devices like printers, CD-ROMs and audio output were also being added to the computers to make them versatile.

Much before personal computers had been developed, a US government agency, Defense Advanced Research Projects Agency, had been financing research in developing a communication system between computers. The idea was to have a distributed computer infrastructure which could survive in case of a nuclear attack. In 1969, the first such system, called the Arpanet, had been commissioned to connect government agencies and some universities. A precursor to the Internet, Arpanet allowed the sharing of data between computers but was very restrictive in its applications. Simple applications like electronic mail and file transfer protocol emerged for users to make use of the network.

In 1973, two researchers, Vinton Cerf and Robert Kahn, developed a protocol for efficiently transmitting information in an error-free way. The Transmission Control Protocol, or TCP, used packets for transmitting information and soon became the standard for all networks. By 1985, many new networks, commercial as well as governmental, had come into being and most networks had shifted to TCP/IP (Internet Protocol). The TCP split the large data files

into smaller chunks or packets and these were then sent by different routes to be reassembled at the destination. The addressing system which allowed this and controlled the routing of the packets was the Internet Protocol. The Internet a conglomeration of various interconnected networks, was gaining in popularity, but even now its use was restricted to large corporations academia and the government.

This changed in 1991 when Tim Berners-Lee, a scientist working at the European Particle Physics laboratory CERN in Geneva, developed a way to share information among the large number of collaborators and scientists at CERN who were working across many continents. He developed a protocol based on hypertext and called it HTTP which enabled the creation of the World Wide Web on the Internet.

In 1993, the National Center for Superconducting Applications (NCSA) released the first web browser Mosaic. This technology was commercialized by a company called Netscape which released the first commercial browser, opening up new vistas to the broad public. The World Wide Web grew exponentially with more and more people using the Internet.

By the mid-1990s, both the Web and the Internet had become indispensable in everyday life. By the end of the century, about 6 per cent of the world's population had access to the Internet. The Internet, or more specifically the World Wide Web, became an important tool for information, commerce, education, communication and virtually all other activities.

Lasers

In 1916, Einstein described the theory of stimulated emission. He proposed that under certain conditions, atoms could be made to emit coherent light—a phenomenon which could either happen spontaneously or under simulation by light. These ideas were mere theoretical curiosities till 1951 when a young researcher, Charles Townes, at Columbia University applied the ideas to microwaves. There had been a lot of development in microwaves during the world

ar and this assisted Townes since some of the echnology was now available. In 1953, he made working device which used ammonia molecules n a microwave cavity to generate very coherent microwaves. The device was called a MASER (or Microwave Amplification by Simulated Emission f Radiation). Independently, around the same me, two scientists at the Lebedev Laboratories n Moscow, Alexander Prokhorov and Nikolai G. Basov, also developed this device.

Throughout the 1950s, there was a lot of research into masers but there were not very many applications where it could be used. Then n 1958, Townes and Arthur Schawlow proposed building a similar device with visible light instead f microwaves. Several groups tried to build this, and finally in 1960, Theodore Maiman built the first laser using a ruby crystal. Six months later, he first gas-filled laser was built by Ali Javan and thers. About two years later, a semiconductor aser had been built.

focused onto a very small area and do not disperse even after traveling long distances, accurate measurements of distance are possible using lasers. For instance, the distance to the moon was measured accurately using laser light reflected from a mirror placed on the surface.

These and other inventions show that the second half of the 20th century was very eventful in terms of developments in science and technology. Human beings increased their reach and went into outer space. Technology was increasingly used for destructive purposes with the development of huge amounts of highly effective weapons. However, technology was also used for development. The electronic revolution led to the widespread use of consumer products in households.

Revolutionary developments in transport and communications narrowed distances between countries though this did not necessarily lead to any better understanding between nations. The developments in the field of information technology

Barcode scanning

Barcodes were invented in the late 1940s but came into use only in the late 1960s and became ubiquitous in the 1980s. They are typically used for identification of items and have proved to be of great use in inventory control and manufacturing. There are different kinds of bar code readers but most of them now use lasers.

The development of the laser was a great technological breakthrough but it didn't find much practical use in its early years. Gradually, t found use in a variety of applications. In 1963, researchers made the first hologram using the newly developed lasers. By the 1970s, bar-code scanners were in use in supermarkets and even libraries. It was also used for eye surgery to fix detached retinas without the need for incision. Compact discs and later DVD technology, which revolutionized the entertainment industry, became possible because of this.

Lasers were used in manufacturing—it made accurate welding or drilling possible. The enormous expansion of fiber optics for communication was a result of this technology. Huge lasers are also routinely used in producing the high temperatures needed for nuclear fusion. Because they can be

were stupendous, especially during the last two decades of the century. The growth of the Internet led to a fundamental change in societies since information was now available easily and thus led to a larger democratization of the marketplace and society.

The rapid pace of technological innovations and their increased use was already putting a huge pressure on the resources available in terms of materials, fuels and even water and air. The environmental impact of several technologies was huge and potentially disastrous. The choice between technological and economic growth and preservation of the planet for future generations became even more stark during the last few decades of the 20th century, resulting in the development of environmental-friendly technologies in the next millennium.

TECHNOLOGY AND ENVIRONMENT

The use of technology for development has always been entwined with its impact on the environment. However, the effect of technology on the environment was not very noticeable in the past. This was because of many factors: there were few people and enough natural resources around, the technology was integrated and developed mostly locally, and the change was gradual, over years and decades. All this changed with the Industrial Revolution.

The Industrial Revolution brought about profound changes in technology, and consequently in almost every aspect of human life. Social relations, economic relations and even the politics were impacted by the massive changes in technology. However, till the 19th century, technology was almost uniformly seen as benevolent—something which increased economic output and hence economic well being of nations, though not necessarily of all its citizens as was evident in the grotesque conditions in the early textile mills in England. Technology

Glacier ice melting

With the increase in greenhouse emissions due to enormous increase in the consumption of fossil fuels during the second half of the 20th century, it is estimated that there is already an increase in the mean temperatures around the globe. Scientists estimate that global warming will lead to a catastrophic decrease in polar ice caps and an increase of the sea level.

was an object of pride and curiosity. The Great Exhibitions of the late 19th century—displaying manufactured goods—were attended by thousands of people who were left in awe of what was possible with machines.

There were of course dissenting voices, most notably the American writer Ralph Waldo Emerson, who even in the middle of all this technophile euphoria, saw technology as overpowering human beings and disputed the idea of conquering nature which had been dominant in Western thinking since the Renaissance. Subsequently, many other influential thinkers like H.G. Wells, Henry James

and Aldous Huxley pointed out the dehumanizing nature of unbridled technological growth. Despite these protests, technology continued to grow unabated. This was as true in the capitalist West as in the socialist East where huge industrial expansion took place after the Revolution. The adverse effects on the environment were brushed aside as a small cost of progress. It was believed that nature was bountiful and robust enough to absorb the industrial waste. This conviction peaked in the 1950s and 1960s, and was sealed with the discovery of nuclear power, the unlimited source of energy.

However, the Great Depression of 1929 took some of the shine off technology when it was realized that technology alone cannot be a panacea for all ills of human society. Socio-economic relations were also an important consideration for human progress. The range of devastation in World War II especially the atomic bombing of Hiroshima and Nagasaki, led to more questioning of technology and its role as a constructive or destructive force in human affairs.

The dawn of the nuclear era led many of the key scientists and philosophers to come out openly against the development of nuclear and other destructive weapons. But the cold war that followed the world war led to a suppression of dissent in the US and there was a silent consensus on the development of technology for military purposes to face the Soviets. The economic boom after the war led to increasing prosperity and a huge growth in consumption in the West, especially in the US.

Though technological growth continued unabated during the 1950s and the 1960s, there was also a growing realization about the impact of technology on the environment. The scientific studies carried out on the impact of various chemicals and industries on human and wild habitats were enough to convince many people that not all kinds of technological advancement is good or sustainable. Several intellectuals and public figures started questioning the capacity of the planet to sustain the hugely resource intensive, technologically driven lifestyles being promoted in the West. The environmental movements were a direct offshoot of this concern with the degradation of the environment. The ecological footprint of an average Westerner was so large that it was feared the sustaining capacity of the planet would be exhausted much before other countries reached that level of development.

Around the 1980s, scientific studies showed that the long-term impact of technology driven growth could be catastrophic. In 1985, British scientists discovered a "hole" of the size of the US in the ozone layer. The huge ozone hole was directly linked to the increased use of aerosols and chlorofluorocarbons (CFCs) in refrigeration and other industries. The ozone layer in the atmosphere protects the earth

From harmful ultraviolet radiation emitted by the sun. A depletion of this layer would mean increased incidence of skin cancer and other diseases. This led to the signing of the Montreal Protocol in 1987, an international agreement to phase out the harmful chemicals. This agreement was highly successful since most countries signed it and the use of CFCs reduced substantially.

The impact of the environmental movements was varied in the West. In parts of Europe, they were very influential in shaping policy. The German Greens, for instance, became a formidable political force with an explicit environmental agenda. Most countries had non-governmental environmental organizations which tried to increase awareness about waste, recycling and hazardous materials.

The impact of these efforts was uneven and by no means immediate. In some countries, legislation was enacted to ensure that industries and individuals do not harm the environment. There were tougher emission standards, stricter rules for storing and disposing of hazardous wastes and a tougher regulatory regime for nuclear power plants. Individuals also responded to this in various fashions—the recycling of glass, paper and aluminum became more common and lifestyles were modified to protect the environment. Companies realized the commercial benefits of being seen as environmentally sensitive organizations, and most of them started running environmental campaigns to offset the bad publicity generated by their core activities.

While this trend was noticeable in the West, even in developing countries, where the issues were of a different nature, there was a growing awareness that the path of development followed by the West was not sustainable. As a result, people started questioning if technological development is worth the cost of irreversible damage to the environment. Movements against building of mega dams and power projects were successful in focusing attention toward the devastating effect of technology on natural resources and the people sustained by these resources.

Around the same time, many scientists started mapping the impact of human activity on the atmosphere, especially the release of carbon dioxide and other products of burning fossil fuels. The accumulation of carbon dioxide and other greenhouse gases—methane, tropospheric ozone and nitrous oxide—causes global warming, or the warming of the global climate. This, in turn, would lead to sea levels rising, polar ice caps melting and other disastrous phenomena.

By the end of the century, though there was no general agreement among scientists about the extent of global warming, there was a consensus that this phenomenon exists and it could be a major threat to the global climate. In 1992, the scientists and policy makers at the Earth Summit in Rio de Janeiro decided to take action on this issue. Five years later, the Kyoto Protocol, an international agreement to curb emissions of carbon dioxide, was signed by many countries.

As the harmful effects of technology on the environment are now evident, there is a call for sustainable development—a system of development for current generations that doesn't compromise the needs of future generations—to strike a balance between human and environmental needs. For this, a more than 50 per cent cut is needed in the global greenhouse gas emission. While the developing countries should be allowed to develop, though with caution, the onus of technological slowdown lies on the developed countries.

Greenhouse gases

The emission of carbon dioxide in manufacturing, transportation and power generation is a major source of the increasing greenhouse effect. Although scientists disagree on the magnitude and pace of increase of global warming, there is a consensus now that it is a real threat to humanity. The disastrous impact of global warming on the livelihoods of billions of people, especially in the developing world, has caused countries to take action to curb the emissions.

Kyoto Protocol

In 1998, an international treaty was drawn up to put into effect measures which will curb greenhouse emissions. The Kyoto Protocol was approved by most countries over the next few years though the biggest producer of greenhouse gases, the US, has not yet signed it. Here David Merrill, Executive Director of the National Global Warming Coalition, reads from a paper urging President George W. Bush to sign the Kyoto Protocol during a protest on the Ellipse behind the White House on February 14, 2005 in Washington DC.

A NEW MILLENIUM

RISE OF CHINA

The 20th century had been a period of unprecedented technological and scientific progress. Of course, there had been revolutionary developments in technology all through human history. Agriculture, domestication of animals, wheel, gunpowder, steam power—these and other inventions in history changed human life in as fundamental a way as automobiles, genetic engineering, computers and space travel. However, two things stand out with regard to the 20th century technological evolution. First, the speed of technological change was much faster than ever before. Secondly, technology became fundamental to the lives of a majority of the humanity. Technology in the 20th century became not only all pervasive but also, in some ways, cheap and accessible.

As befits a century whose defining theme was science and technology, the century ended with a huge technological problem looming on the horizon. This was the so-called Y2K problem which threatened to create mayhem and chaos throughout the world. The genesis of this problem lay in the

huge mainframe machines which were used b most corporate and government agencies to sto and process data. These machines, which we popular before the big microprocessor revolution had been in use (with minor changes) for a ver long time since the cost and effort involved migrating to newer machines and programs wa very large.

The problem was that most of these machine which were at the heart of the modern industri society, stored dates in such a way so as to indicat only the last two digits of the year rather tha the full four digits. When these were coded in th 1960s, no one thought that the programs and th data would last till the year 2000. So there was a lo of apprehension that the calculations of time don by these programs to determine things like annui and interest payments would all go wrong at th turn of the millennium, as it would assume the yea 2000 to be 1900! Therefore billions of dollars an many man years of computer professionals wer spent in trying to avoid this problem. Finall when the clock struck midnight on December 31 1999, everybody waited with bated breath for

A depiction of NASA's X-43A, Hyper-X research aircraft

Hypersonic aircraft, traveling at more than five times the speed of sound, has been a goal of aviation engineers for a long time. However, there are many technical issues which need to be solved before this becomes possible even at the prototype level. In 2007, the US Army announced the successful test of NASA's X-43A—a hypersonic, scramjet-powered research aircraft—designed to fly at 10 times the speed of sound. Scramjets, using oxygen from the atmosphere—unlike rockets which carry oxygen onboard—may reduce the travel time from San Francisco to Washington DC to 20 minutes (which is about 7 hours now).

2000: George Bush becomes the president of the US and Vincent Fox becomes the president of Mexico. The full genomic sequence of the flowering plant *Arabidopsis* is published. A preliminary draft of the complete genome sequence of humans is completed by the Human Genome Project.

2001: Ariel Sharon elected as the prime minister of Israel. Milosevic surrenders in former Yugoslavia and is tried for war crimes. Terrorists attack New York, Washington and Pennsylvania by blowing up planes. The World Trade Center attacked by terrorists, and over 3,000 people killed. Terrorist attack on the Parliament of India. The US and allies attack Afganistan and take over Kabul from the Taliban. Gaur, an endangered species, is cloned for the first time. The first artificial heart is implanted into a patient. Apple Computer releases the iPod, a portable music player which would revolutionize the entertainment industry. Microsoft unveils its own game console Xbox while Nintendo releases the GameCube to counter Sony's Play station.

2002: Hu Jin Tao takes over as General Secretary of the Communist Party of China. Many countries in

Europe switch to a common currency, the Euro. The US Mars Odyssey probe begins its imaging of the Martian surface. The probe finds huge deposits of ice on Mars.

2003: The US led coalition invades Iraq, toppling Saddam Hussein's regime. Saddam is captured by the US forces. Vladimir Putin re-elected president of the Russian Federation. SARS, a deadly respiratory condition possibly linked to Avian Flu, spreads over parts of East Asia. The Human Genome Project ends with a complete genomic blueprint of human beings. China launches a taikonaut into space.

2004: European Union expanded to include many countries of the erstwhile Soviet Bloc. George Bush re-elected resident of the US for a second term. A huge undersea earthquake near Indonesia triggers a devastating tsunami which kills hundreds of thousands in Asia. US probe lands on Mars and investigations confirm the presence of water sometime in the past on Mars. Scientists in South Korea announce successful cloning of human embryos. SpaceShipOne, the first privately funded spaceflight, is launched. Cassini-Huygens spacecraft arrives at Saturn to study its satellites and the rings.

2005: Pope John Paul II dies and Pope Benedict takes over as the 265th pope. Deep Impact, a probe launched by NASA, crashes into the Comet Tempel 1. Huygens lands on Titan, the largest moon of Saturn. Kyoto Protocol, a global agreement to curb emissions of greenhouse gases, comes into effect.

2006: Saddam Hussein executed after being convicted in Baghdad. NASA's Stardust mission brings back dust from a comet for investigations.

Scramjet, an airplane to fly at seven times the speed of sound, is tested. The first railway to Lhasa in Tibet is opened.

2007: Gordon Brown takes over from Tony Blair as the prime minister of Great Britain. Nicolas Sarkozy elected President of France.

Sunetra Sunray, a prototype two-person solar-electric vehicle designed in the 1990s on the island of Hawaii

nancial and administrative catastrophe to occur. As it turned out, nothing serious happened and so the next millennium began on a good note.

The events in the closing years of the previous century determined the overall framework of the state of affairs in the first few years of the 21st century. The breakup of the Soviet Union and the chaos that followed in eastern Europe had left the world with one unmatched military and economic superpower, the United States. However, the last few years had seen the rise of a hitherto underdeveloped country into a potential giant. China had started on a path of economic reforms which led to its economy growing at a blistering pace. The sustained rate of growth of the Chinese economy over the last 15 to 20 years is unmatched in the history of humankind. This, coupled with the fact that China is home to more than 15 per cent of humanity, has had profound consequences for the whole world in the last few years.

An important trend in the world economy has been growing globalization and interconnectedness of the different countries. With the innovations in communications, especially the Internet, geographical boundaries are less and less relevant. Thus nations seem to be increasingly using their competitive advantage to trade among themselves. China has become the workshop to the world where most of the consumer goods are produced. The supply chain management and the logistics required for such a global sourcing of goods are quite challenging. However, with the extensive use of communication technology and innovations like container ships for quick transport, this is now a reality. It was not just manufacturing of goods which became global, communications enabled the financial sector to globalize even faster and to an even larger extent.

The growing globalization also had an impact on innovation. With growing competition from around the world, innovation became an important element for any industry that was to survive. The innovations—in hardware, software, or even business processes—could lead manufacturing processes to greater efficiency, resulting in cheaper and better quality goods. Therefore innovations became the distinguishing factor among organizations and, to a large extent, decided which companies and industries survived in the current economic system.

The US, which had emerged as the unquestioned leader in technological and scientific innovation, continued to maintain its lead over Japan and Europe. The research universities in the US continued to be the powerhouses of ideas and inventions in all fields. The global corporations,

some of whose sizes surpassed the economic size of many nations on earth, were also spending huge amounts of finance and infrastructure on research and development. A new phenomenon of venture capital also allowed small groups of individuals with innovative ideas to turn them into reality. This trend was particularly true in the fields of biotechnology and information technology.

The 21st century is still less than a decade old. Though there have not been too many significant inventions in this short period, there has been ongoing research in a variety of fields which has laid

the foundations for many innovations in the coming years. A survey of the trends in various fields will thus give an idea of what kinds of inventions might be possible in the near future. The survey shall, of course, talk about the innovations in the 21st century that have already taken place, but will focus on the trends in research and development in the areas of food and agriculture, communications, information technology, materials, transport, energy and medicine.

Food and Agriculture

The last decade of the 20th century saw a huge increase in the adoption of transgenic crops by farmers. The increase in yields, and also their herbicide and pest resistance were important considerations for commercial farmers. In fact, the technological diffusion seen in the adoption of these genetically modified (GM) crops was unprecedented in the history of agriculture. In fact, by 2006, GM crops were being grown in 20

Chinese exports

The emergence of China as the "workshop of the world" is one of the major developments in the 21st century. Cheap labor, good infrastructure and good logistics have made most companies shift their manufacturing base to China. China is the world's biggest toy exporter, selling 22 billion toys overseas in 2006, which is 60 per cent of the world market. However, there have been some issues with labor practices and also the use of toxic and harmful materials in manufacturing recently.

Genetically modified (GM) food

Genetic modification in food, based on the insertion of genes rather than the production of hybrid plants, has become increasingly important in the last decade. There are attempts to develop GM varieties of wheat and rice, which are the main food crops for a majority of the population. Here a researcher is inspecting a GM rice plant or "Golden Rice" developed at the International Rice Research Institute (IRRI), Philippines.

Biotechnology greenhouse

The application of biotechnology to agriculture has led to the introduction of many new varieties of crops which are herbicide and pest resistant. Cotton, soybean and maize are among the plants for which many GM varieties are available; these varieties are very successful commercially. Though there are fears about the replacement of natural, ecologically safe varieties by GM plants, there has been an enormous increase in the acreage sown with GM crops. Here researchers are seen carrying out trials of GM crops in a greenhouse.

countries on more than 100 million hectares. The major GM crops are soybean, cotton and maize or corn, where more than one half of the soybean being grown in the world is genetically modified.

Identifying commercially useful plant genes has been the focus of agricultural research in the past few years. These genes could then be transferred into other plants to give them desirable properties. For instance, in 2006, researchers in Australia discovered a group of frost resistant genes in a particular type of grass which grows in Antarctica. This grass is able to withstand temperatures down to −30°C because of a protein which does not allow ice crystals to grow. This gene, if it could be transferred into wheat and barley, could prove to be immensely useful. Similarly, a rice gene variant has been identified which will allow rice plants to be completely submerged in water for over two weeks without damage. If this gene variant could be transferred to the commercially used rice varieties, it could prove to be a boon for rice producing countries.

The introduction of transgenic varieties of many more plants would continue. These plants would be more resistant to pests and thus reduce the need for pesticides which contaminate the water and soil. There have been some attempts at introducing genes into rice and other cereals to increase the amount of micro-nutrients and essential vitamins in them. If this could be replicated at a commercial level, it would go a long way in reducing malnutrition and morbidity due to deficiency diseases.

In the coming decades, there will be much research on producing super-rice varieties as well as wheat varieties with higher yields, leading to an increase in productivity of over 15 to 20 per cent.

The strategy followed in agricultural research will continue to rely on biotechnology, conventional breeding methods and also better farm practices to increase yields. It is important to increase yields as the agricultural land cannot increase to keep pace with the growth in population and its needs. Although the world produces enough food for every person on earth, increasing meat consumption worldwide may lead to ecological imbalances which will need to be addressed.

Water resources will prove to be a major constraint in agriculture. Although more than 70 per cent of the planet's surface is water, only about 2.5 per cent of it is fresh water, most of which is locked in the polar caps and in very deep aquifers. The water available for human use is in the rivers, lakes and accessible aquifers. A large part of it is used in agriculture, and thus technologies which use water efficiently will be important in the future. Drip irrigation is already being used extensively worldwide and there are also other technologies under development which would provide water to plants only when needed.

Research of leguminous plants like peas and beans has shown that it might be possible to coax non-leguminous plants to fix atmospheric nitrogen. Leguminous plants form nodules on their roots with the help of rhizobia bacteria. The bacteria fix the nitrogen into a usable form which is imbibed by the plant. This is why alternating cereals with legumes is useful in restoring soil nutrients. Scientists have succeeded in mutating a gene in legumes to produce nodules without the aid of the bacteria. This, if successful in plants like rice and wheat, might reduce the need for artificial fertilizers and increase yields substantially.

Though many animal genomes have been decoded completely, only three plant genomes have been completely decoded so far. These include *rabidopsis*, rice and black cottonwood poplar. Researchers are hoping that a better understanding of the cottonwood poplar genome would allow them to improve the cellulose produced by this fast growing tree. If this is possible, then the tree could be used as a source of cellulose which could be converted into ethanol for use as fuel.

Energy

In terms of energy use, the fossil fuels were the key drivers of economic growth in the 20th century. The increased use of the automobile and aircraft for transportation meant a huge increase in the consumption of oil. Similarly, electricity consumption increased dramatically, the production of which depended mostly on coal or gas fired power plants. Although the developed countries used a disproportionately larger share of energy than developing nations, the spurt in the economic growth in countries like China and India has led to their using large amounts of energy and fossil fuels. The energy use in the 21st century will increase substantially as the developing countries embark on a path of energy-intensive growth which the developed countries have followed.

Although new deposits of fossil fuels are being found regularly, there is still only a finite amount of it in the earth's reserve. In the last few years, as the more productive wells have dried up, technology has made possible to squeeze more oil and gas from hitherto uneconomical wells. The development of technology to extract oil from tar sands has also been of significance. However, these technologies can only delay oil and gas exhaustion, they cannot prevent it. Ultimately, humans will have to look at alternative sources of energy to power their automobiles, machines and homes. The future has to belong to renewable sources of energy not only for sustainability but for environmental reasons.

The increasing emission of greenhouse gases by human activity has led to a situation where it is widely recognized that climate change is inevitable unless drastic measures are immediately taken to control the emissions. There have been several international initiatives on this front, including the Kyoto Protocol, but none have resulted in decreasing the emissions substantially. Renewable sources of energy like solar, wind and hydrogen power could prove to be of importance in controlling the fatal emissions.

Solar cells or photovoltaics attempt to convert the energy from the sun directly into electricity at the atomic level, using the photoelectric effect. This technology, first discovered by the French physicist Edmond Becquerel in 1839, was a subject of scientific curiosity for several decades. It was with the development of the semiconductor industry in the mid-20th century that the first steps toward the commercialization of photovoltaic cells were taken. As the Czochralski process was developed for producing pure crystalline silicon in the 1940s and early 1950s, it was realized that semiconductor technology could be used to build efficient solar cells. In 1954, Bell Labs developed a crystalline silicon photovoltaic cell, with an efficiency of 6 percent.

Solar cells were mostly used for specialized applications throughout the second half of the 20th century. Space probes and satellites used solar cells to meet their power requirements. In addition, these were used in areas where the electric grid was not available and laying of power cables was prohibitively expensive or difficult. But they have not been hugely popular for day-to-day needs because the per unit cost of electricity produced by solar cells is much higher as compared to traditional sources of energy.

However, in the last few years, many design and manufacturing innovations have drastically reduced the cost of photovoltaics and increased their efficiency dramatically. These include the use of novel materials to make solar cells as well as the introduction of thin film solar cells. Thin film solar technology can be used in a variety of situations— for instance roofing material, which can serve as a solar generator to produce power. In addition, research on concentrating solar energy has led to increased efficiency of power generators, resulting in lower costs of solar power.

Apart from solar cells, solar heaters—used for heating buildings and water—also saw major improvements, including the better designs and new materials to improve efficiency.

Solar power tower

There have been several attempts at using solar energy for generating electricity. However, till recently, photovoltaic cells which were used for this purpose were too inefficient, leading to a much higher cost of electricity. In the recent years, there has been tremendous development in semiconductor technology, increasing the efficiency of solar photovoltaic cells enormously. This solar tower in Seville, Spain supplies power to up to 6,000 homes.

Wind power

Wind power, used for centuries for mechanical work, has been used for generating electric power only in the last few decades. The wind turbine design has undergone many modifications and now there are turbines which can produce up to 1 MW of power. Wind farms, onshore and offshore, have gained in popularity and become a common sight. A row of wind turbines in a field near Columbia River in Oregon, USA, is shown here.

Fuel cells

Fuel cells, which had been hitherto only used in remote locations and on spacecraft, are increasingly being considered for a variety of applications, especially in transportation. Hydrogen and fuel cell technology is becoming popular due to its green appeal—there are no moving parts and no burning of fuel involved. There are still several design and cost-efficiency issues which need to be resolved before they can find extensive applications. A prototype of a micro hybrid hydrogen storage device is shown at the International Hydrogen and Fuel Cell Expo in Tokyo, 2007, where over 500 companies participated.

Wind is another source of energy which has not yet been tapped fully. Although the use of wind power has increased in the last few years—wind power generation increased four times between

2000 and 2006—it still only provides less than 1 per cent of the global electricity. However, the recent focus on global warming has led many nations to promote wind power as an alternative, green source of energy.

Most modern day wind power generators use wind power to move turbines which are connected to electric generators. The electric power generated can then be used locally or the power can be fed into the grid. Recent advances in turbine design has made generation with even moderate wind speeds a possibility. The use of offshore wind farms is another development in recent years. Here, large turbines are installed several kilometers from the shoreline and the power generated fed into the grid. There are many advantages of offshore wind farms. Being close to cities, they save energy because of transmission lines. Moreover, they make use of high wind speed in open seas, resulting in a more efficient turbine.

The advances in turbine design has led to increased power generation, making it cost effective. The largest turbines in use now are producing several megawatts of electricity and this is expected to increase in the future. The cost of wind power, most of which is from the capital cost of setting up the generating capacity, compares favorably with the cost of generating power using coal or natural gas.

The potential for wind power is huge since the operating costs are minimal and the production of power does not produce any emissions, unlike conventional methods of electricity generation.

Even if a small percentage of the available wind energy is tapped to produce power, it could make a substantial difference in the greenhouse emission as thermal power plants are a major source of greenhouse gases.

Geothermal and tidal power are the other sources of energy which are likely to see growth in the coming years. Although these are not very popular at the moment, except in localized pockets like Iceland and in France, there is a growing interest in these renewable sources in many countries.

Fuel cell technology is another area where many developments have taken place in the recent past and the future holds promise. Fuel cells convert chemical energy directly into electrical energy. They are quite like batteries, except that in fuel cells there is usually a mechanism to replenish air and fuel so as to make them much more longer lasting than batteries.

Fuel cells in their basic form have been around since the late 19th century but their use really started with space travel. Spacecraft and satellites needed reliable sources of power (other than solar power), and hence, during the 1960s, a lot of development took place in making smaller, more efficient and longer-lasting fuel cells to power spacecraft. In recent times, the need to control emissions from fossil fuels has led to the use of fuel cells in commuter vehicles as well.

Fuel cells are of various types, but the most common ones in recent times have been hydrogen fuel cells. These use hydrogen and oxygen with the various combinations of catalysts. The last few years have seen tremendous efforts in developing cheaper and more efficient catalysts. Since the catalyst cost is a major component of the cost of making a fuel cell, better and cheaper catalysts have the potential of making fuel cell economically viable. For instance, in 2002, the typical catalyst cost for producing 1 kw of power was around $1000, which dropped significantly in later years.

In recent years, there has been a lot of interest in hydrogen-powered vehicles. These vehicles have

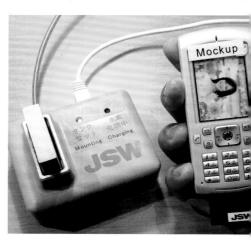

he potential of generating no pollutants since he waste product is expected to be only water. owever, there are several issues which need to be ddressed before they can become economically iable, including the costs of fuel cells using ydrogen and the production and storage of ydrogen on a large scale. Since hydrogen is not vailable in pure form on earth, it will need to e extracted from hydrocarbons, mostly natural as or methane. Other methods could include irect production by electrolysis of water or direct hemical production from hydrogen compounds. s energy is needed for the conversion of methane ito hydrogen and for the electrolysis, it will result i some emissions and also reduce the net efficiency f hydrogen cells. The direct chemical production

are solved and a hydrogen network established, it is very likely that the internal combustion engine will have its first real competition in more than a century. Cheaper fuel cells can be used not only for generating power in locations where the grid is not available but also to supply power to the grid once the costs of manufacturing come down. With the renewed interest in alternate energy sources due to the global warming caused by greenhouse gases, many governments are encouraging the development of fuel cells for the purposes of power generation and transportation.

In the long term, nuclear energy has the potential of meeting the energy requirements of the earth on a sustained basis. Nuclear energy, which currently only relies on nuclear fission for generating power,

Alternative technologies in transportation

The automobile companies, because of pressure from the regulatory agencies and also from the consumers, are spending a lot of money in the development of alternative technologies to the internal combustion engine which has been the mainstay of personal transportation for over a century. Fuel cell technology has evolved to a point where it is only a matter of a few years before fuel cell powered vehicles could be competitive to petrol burning ones. Here General Motors Europe President Peter Forster is posing with a GM concept car at the Geneva International Motor Show in 2007.

ising catalysis has the advantage that no external nergy is required to be pumped in.

In the last few years, several innovations have aken place to make hydrogen vehicles a practical eality in the near future. These include safer torage methods onboard, cheaper methods to xtract hydrogen from water, as well as extracting ydrogen cheaply from sugar and water mixtures. here are already prototype vehicles which are unning on hydrogen fuel cells, though in the bsence of a reliable and widespread hydrogen conomy, they cannot become commercial reality. Several automobile companies are planning to elease hydrogen vehicles in the near future while rototypes of boats and motorcycles also exist. Boeing is experimenting with an aircraft powered y fuel cells and batteries.

Once the problems associated with the production, storage and transportation of hydrogen

faces many technological challenges currently. These include safe disposal of long-lasting waste products as well as accident prevention. The mishaps at Three Mile Island in the US and Chernobyl in the former Soviet Union reminded the world of the hazardous effects of nuclear energy. Unless these concerns are addressed by better technology and new innovations in reactor technology, nuclear power in its current form will never really acquire public acceptance and widespread usage.

Fusion energy has been a long-standing goal of energy scientists. Nuclear fusion, unlike fission, is a very clean form of energy as it produces no radioactive waste material. However, despite intensive efforts by scientists over the last five decades, the goal of achieving economical fusion energy remains distant. The technological challenges posed by fusion are very formidable and any breakthrough in the near future is difficult.

Transport and Communications

The 20th century brought about revolutions in transport technology on land, in air and on sea. Land transport saw the advent of the internal combustion engine which underwent many modifications and improvements by the end of the century. The design of railways, where the original steam locomotive was almost totally replaced with diesel engines and electric power, had been innovated to increase the speed and improve comfort and safety.

Air transport saw many revolutions since the invention of heavier-than-air machines by the Wright brothers in the beginning of the 20th century. Bigger and faster airplanes were developed in the last hundred years. The invention of jet propulsion and the subsequent improvements in technology led to the tremendous growth of air travel to the extent that air travel has now almost replaced passenger travel by sea and railways for longer distances. Supersonic civilian and military aircraft became a reality, as did transport aircraft capable of carrying huge loads of cargo.

There were many improvements in the design and technology of ships, submarines and other sea transport. Steam power was replaced by diesel and electric power in ships and submarines, and even nuclear power in some military transport. Ships became bigger, faster and safer with improvements in design and electronics. In fact, in all forms of transport, the introduction of computers and electronics after the 1970s was a major innovation. Improved communication and navigation techniques, including the use of satellite navigation, made transport much safer. Innovations in new light weight materials like carbon fiber and composites also led to bigger and better designed aircraft, passenger cars and ships.

Space exploration, one of the crowning technological achievements of the late 20th century, also saw major innovations since the time of the Sputnik. By the end of the century interplanetary probes were common and they were equipped with better imaging technology for exploration of distant planets and other objects. The International Space Station or ISS was well on course with regular space flights to equip and enhance it into a platform in space for farther travel and experimentation.

The 21st century is likely to take forward several innovative ideas in transportation technologies. In the field of automobile technology, hybrid cars with the internal combustion engine and electrical propulsion will become more common. This will depend to a large extent on the efficiency, affordability and design innovations in fuel cell technology. Cars, which are already heavily automated in their engine components, will see further incorporation of microprocessors and electronics to make them better and safer. Although it is difficult to see any technology in the near term totally replacing the internal combustion engine, hybrids or pure electric vehicles will certainly become more popular if the necessary economic incentives are provided.

In railways, high-speed, magnetically levitated trains will become more common. These trains travel at very high speeds because they do not touch the tracks while in motion. This is achieved by producing a very high magnetic field between the tracks and the locomotive. With the development of high temperature superconductors and their incorporation into bulk structures like rails and wires, magnetic levitation would become more affordable and hence more widespread.

Aircraft will also see many innovations. Supersonic transport is already a technological reality, though there are environmental issues related to it in terms of noise pollution and greenhouse emissions. Hypersonic transport, which will fly at more than five times the speed of sound, is another possibility. The most

Magnetically levitated train in Germany

Using magnets for propulsion of trains on rails is an old concept which did not find practical application because of the lack of appropriate technology. In the past decade, the technology for creating and sustaining the enormous magnetic fields has been developed, and now magnetically levitated trains are a reality.

SpaceShipOne, a private spaceflight

In 1996, the X-Prize was instituted to
reward the first privately developed
vehicle which could fly to an altitude
of 62 miles (100 km), the edge of
outer space. In 2004, SpaceShipOne,
developed by Scaled Composites, won the
$10 million prize. Here Pilot Michael
Melvill is seen atop SpaceShipOne after
successfully completing its test flights in
the Mojave Desert.

promising technology in this regard is the scramjet
technology. Scramjets are a variation of the basic
jet propulsion technology which uses supersonic
airstream to generate very high thrusts. Scramjets
need a minimum speed to function and hence
require an auxiliary technology to propel them
to supersonic speeds. There have been several
successful experimental scramjet projects, though
the technology remains controversial given the
high costs and the complexity involved.

Outer Space

Space exploration and travel is also bound to be
revolutionized by technology in the current century.
While the national space agencies like NASA, ESA
and RSA are focusing on the International Space
Station and on sending interplanetary probes for
expanding our knowledge about the solar system
and its constituents, private agencies too are
getting into the fray.

There are already plans for a space hotel or
the use of a space station as a hotel. Several
experiments have been made with an inflatable
space habitat which could serve space tourists.
Some individuals have already flown into space
in their private capacity and this trend is likely
to grow. In terms of innovation, the X-prize, a
$10 million prize offered to a private company
which could design and make a vehicle capable
of reaching an altitude of 62 miles (100 km)—
the boundary of outer space—has already been
claimed in 2004 by SpaceShipOne. The America's
Space Prize of $50 million for a US company to
make a reusable space vehicle to carry passengers
and cargo to the space station has been announced.
Several companies are already planning suborbital
flights to an altitude of about 62 miles (100 km) to
take individuals into space to experience the thrill
of space travel.

Communications

In the field of communications, the growth of mobile
telephony in the first few years of the 21st century
has been phenomenal. There are now almost twice
as many mobile phones as fixed telephones in the
world and this gap is bound to grow since, unlike
fixed telephones, there is a huge potential for growth
in the mobile telephone sector. New technologies
in mobile telephony, including the so-called 3G
technologies, will become more widespread.
These technologies will allow much faster data
transfer speeds and thus make possible mobile
Internet, video downloads and other applications.
Voice over IP, a way to use broadband Internet
to transmit voice calls, is another area which will
grow since the cost of telephony in this method is
significantly less than traditional methods. Better
compression technologies will allow very high
fidelity transmission of audio and video signals.

WiMAX (Worldwide Interoperability for
Microwave Access) is a technology which has the
potential to revolutionize communications in the
near future. This technology provides wireless
data over long distances and this can be used for
telephony and Internet access. Since the range
of WiMAX is much larger than the current WiFi
technology, WiMAX networks can be deployed
over whole cities. This convergence of voice, data
and video will prove to be a boon for the global
areas where access is still limited.

Mobile computing

With the advent of various wireless
technologies, connecting the computer
to the Internet has taken a quantum
leap. WiFi is now ubiquitous inside
buildings and in campuses. The next
generation would be WiMAX for longer
range connectivity within cities and even
beyond. The photograph shows Sean
Maloney of Intel unveiling a WiMAX
enabled mini-tablet Ultra-Mobile PC at
the 2006 Intel Developer Forum.

Weird Gizmos

Gadgets—devices which have some practical function, being normally a clever application of available technology—have played an important part in history since its beginning, though they were not known as gadgets in ancient times. The Stone Age hunters made hand tools to kill animals, which were also a type of gadget. The plow, the wheel, clay pottery, the abacus, the lever, the wine press, and almost all the inventions could be considered as gadgets of their times.

While most gadgets and gizmos are immensely useful to humans, there are some which, though useful in a broad sense, are definitely not ordinary—they use technology in an extremely ingenious way—and therefore can be described as unusual or weird. There must have been weird gadgets throughout history, though we don't have much written or pictorial record of this. It is next to impossible that the fertile human imagination and technical ingenuity would not have created weird gizmos in the past.

In the Golden Age of Islamic Science, the famous Islamic engineering genius Al-Jazari created many clever mechanical contraptions like the automaton used for entertaining royal guests. This was a boat with four automatic musicians who played programmed instruments. Of course, the programming was mechanical and not computerized—with cogs and wheels and cams. Some of Jazari's inventions were reinvented much later and turned out to be immensely useful. For instance, Jazari used a flow regulator in one of his

mechanical contraptions. Many centuries later, the flow regulator was a very important part of the boiler in the steam engine, though Jazari was not credited with the invention.

The Chinese were also very adept at making strange, seemingly useless contraptions. In the third century AD, Ma Jun, a brilliant mechanical engineer, made a hydraulic powered mechanical puppet theater for the Emperor's court. This puppet

theater was fashioned on a wooden wheel which rotated with flowing water. The wheel had many puppets who moved, played musical instruments and even threw swords! Another inventor, Qu Zhi, made a "rat's market" with automatically closing doors and moving figures. He also made a doll's house with bowing figures and automatically closing doors.

The Middle Ages in Europe, the Renaissance, the period of the Industrial Revolution and later, all had ingenious inventors who developed weird gadgets and inventions. For instance, a lip clip was invented in 1924 to get a bow-shaped upper lip without undergoing surgery. Or, a car bib was patented in

Rhino

Equipped to drive on land and sea, the Rhino is making its way up a steep incline.

Spaghetti aid

Devised by amateur American inventor Russell E. Oakes, the spaghetti fork winds up spaghetti strands, making them easy to eat.

he US in 1980 to catch all the food that drivers drop while eating in the car. The fascination with gadgets has a long history—till the advent of mass manufacturing of goods in recent times, these were toys for the amusement of the rich.

The 20th century saw mass production and a relative democratization of the market for goods. In more recent times, after the introduction of electronic devices and computers, most of the weird inventions started making use of electronics and microprocessors rather than mechanical gears and wheels. But there are still enough mechanical gizmos which are ingenious enough to be patented, though their use could be debatable. For instance, a US patent was granted in 2000 for a "Cry No-More" contraption. This was a baby soother which had straps that could be put on the ears of a child to ensure that the howling infant does not remove the soother and start yelling. Or a light bulb changer, an extremely complex mechanical device with springs, gears and motors, that could exchange a light bulb. This gizmo was patented in 2004 in the US. Earlier, in 1976, the Pogo shoes, which combine a unicycle, a pogo stick and shoes to make a dangerously exciting contraption, were patented.

Other whacky inventions include a motorized ice-cream cone which rotated the ice cream for the consumer; an alarm fork which tells the user when to start and stop eating; a doggie umbrella that fully covers a dog with air holes in front; a kissing shield, made of a thin latex membrane, to ward off the risk of infections while kissing!

Though most of these weird inventions are clearly meant more for the amusement and gratification of the inventor, there are some which could actually be very useful. An insomniac's helmet patented in 1992 is a helmet fitted with pads which can move with electromechanical parts. The pads can provide the wearer a gentle head massage and help relieve stress which could be causing sleeplessness. Similarly, a diaper alarm, patented in 1980, is a small device with sensors which could give a visual or an audio signal if the diaper is dirty. A fingertip toothbrush—a rubber cap with bristles which can fit on a fingertip and allow one to brush one's teeth—was patented in 1999 in the US. These and many more inventions, if actually produced beyond the prototype and backed with proper marketing, could prove to be big commercial successes.

Indeed, there are actually some whacky inventions which actually get manufactured and sold. In the aftermath of September 11, 2001 when there was a huge phobia of biochemical attacks, a bulletproof bed which offers protection from biochemical attacks, bullets and explosives, kidnapping and other mishaps was being sold in the US for $160,000. A nose-pouch, a handkerchief with a pouch to contain the nasal discharge, is also available. For the adventurous, there are pierced glasses which use a pierced stud on your face to hold the spectacles, thus doing away with the earpiece which always is either too loose or too tight!

Waterproof cigarette

A woman smoking a waterproof cigarette under a running tap is seen here.

Then there are weird inventions which cater to the specific needs of a group of people. The Islamic car fitted with a device which can tell you the direction of Mecca, or the Kosher mobile phone which blocks most of the pornography numbers and also automatically makes using the phone on Sabbath prohibitively expensive, belong to this category of weird inventions.

Though the list of weird inventions would turn out to be very long, there is possibly one which could come close to being the weirdest invention which was actually granted a patent. In 1975, the US Patent Office issued a patent for a method of combing your hair— the method involves a three-way combing of hair in such a way that the bald patch in the middle could get covered! The layers of hair need to be set with a hair spray, a technique which could be popular with balding men.

Looking at the trend of these whacky inventions, it seems that they are here to stay. Despite the tremendous pressure of society on individuals to conform to the well-trodden paths, there will always be people with an inventive spirit and a desire to come up with ingenious ideas and innovations which would be considered crazy, whacky and weird.

Internet

Internet will continue to grow exponentially as in the last few years. The strain on the physical infrastructure of the Internet is already a major problem and this will become a tremendous bottleneck in the near future. This is because the number of devices accessing the Internet would increase hugely in the next few years. Already, in some countries mobile phones are competing with computers for Internet access. As mobile phones become more sophisticated and their networks become capable of faster data transfer, especially video transfer, the strain on Internet resources could reach a breaking point. As predicted, if devices connected to the Net become ubiquitous (from small embedded chips in consumer appliances and cars), it doesn't seem likely that the current logical and physical infrastructure of the Internet will be able to take the load.

In terms of the logical structure of the Internet, the old naming protocol is already being replaced by a new protocol which will allow many more devices to be on the Internet. The physical infrastructure is much harder to replace, and unless some innovative new technology is introduced, it will continue to lag behind what the needs of the users are going to be.

Computers

Computer technology will continue to improve with better and more powerful processors being developed. Already, the manufacturing techniques in the recent years have reached such a degree of sophistication that it is now possible to embed many more devices on the processor. The increase in density of devices on the chip and the reduction in costs roughly have been following the famous Moore's Law of the number doubling every 18 months. However, unless there is a radically new technology, the miniaturization will reach physical limits. Already, the generation of heat at such device densities is a technological challenge.

The advent of dual-core, quad-core processors in the last few years has added tremendous processing power to the computers. Faster memories have been developed recently, and this trend will continue. Another area where tremendous progress has been made is in data storage. With new technologies it is now possible to pack in much more data onto magnetic hard disk drives than before. Storage, using optical media like high density DVDs, is already a reality, and it is expected that storage technologies will keep on improving.

A very promising area in the field of computers would be quantum computing. Though the concept of quantum computing is quite old, it was only in the 1990s that there was a practical realization of the idea, though on a very limited scale. Quantum computers have the potential to solve some of the most difficult problems in computation, with tremendous applications in various fields like logistics, biology and telecommunications. Unlike the traditional computer where the information is stored in bits, in quantum computers it is stored as qubits (quantum bits), allowing quantum mechanical phenomena, such as superposition and entanglement, to perform operations on this data. The experiments with various technologies, such as the semiconductor technology, over the last few years have shown that it is possible to make quantum computers a practical reality.

Medicine

The developments in medicine in the 21st century will almost certainly eclipse the tremendous progress made in health and medicine during the previous century. The 20th century had seen some of the most revolutionary developments in the science and practice of medicine, developments which had reduced mortality and increased life expectancy for billions of people. These included effective medicines and vaccines against most diseases, very accurate diagnostic methods and techniques, and a better management of public health. By the end of the previous century, it was clear that there would be a qualitative change in medicine in the next century.

Stem cell research had been going on for several years in many countries by the close of the 20th century. Stem cells are undifferentiated cells which have the capacity to produce different kinds of cells needed by the living body. Involving cell replacement in the human body, stem cells have the potential for treating congenital, developmental, or degenerative diseases which involve tissues or organs that do not have the capacity for self-repair. The stem cells are first differentiated into the cells of choice and then transplanted in patients to replace damaged tissues.

Stem cells are of two kinds—the embryonic stem cells which are obtained from the embryo in its early stages of development and adult stem cells which are obtained from the tissues of an adult organism. Embryonic stem cells are especially suitable for cell therapy for two reasons. One, as they are derived from early blastocysts, they have the flexibility to develop into any of the 200 cells that make up the human body. Two, they have the ability to multiply indefinitely. In 1998, James Thomson and his team isolated and created human embryonic stem cell lines at the University of Wisconsin-Madison in the US.

However, by 2002, stem cell research had become very controversial in many countries, especially the US. The debate involved researchers, theologians, politicians and general public. At the heart of the controversy was the fact that starting a stem cell line involves destroying the human embryos and therapeutic cloning—this was unacceptable on ethical and religious grounds to several groups. The US banned federal funding for any embryonic stem cell research, while several other countries put some restrictions on such research.

On the other hand, adult stem cells have a huge potential in medicine. The development of reliable techniques to obtain and study adult stem cells have proved to be a benefit for the researchers. Adult stem cells from various tissues of the human body (such as the small intestines, the skin, bone marrow) have been developed in various laboratories. Host-derived adult stem cells, that is, those which are derived from the tissue of a patient, can be used to regenerate injured organs. For instance, if a patient has a spinal cord injury, stem cells obtained from other tissues of the patient could be harvested, grown in the laboratory and then grafted into the injured portion, where they

will differentiate into the kind of cells which are injured. Similarly, patients with diabetes, cystic fibrosis and other ailments could be cured with appropriate stem cell grafting.

Another area which would be important in 21st century medicine would be gene therapy. Gene therapy refers to the incorporation of a person's genes into cells and tissues to treat diseases. Though the techniques for gene therapy are still not very advanced, substantial progress has been made in the last few years. Researchers have already been able to insert genes into the brain (thus overcoming the blood-brain barrier which doesn't allow foreign bodies to enter the brain), treat some forms of melanoma using killer T-cells, as well as treat inherited retinal disease. In addition, techniques have also been developed which will

Stem cell research

The stem cell research has become a controversial subject and evoked very strong reactions from politicians, ethicists, as well as from several religious leaders. Fears about playing with nature at its most fundamental level as well as public opinion have prompted very strict regulation of stem cell research in several countries, most notably the US.

Human Genome Project

The Human Genome Project, a multi-national collaborative project to map the complete genome sequence of human beings, was completed in 2003. It was a landmark project which will allow scientists and researchers to determine the genetic causes and treatments for many of the diseases and disorders. Celera Genomics Inc was a private company which carried out a similar project.

Insulin inhaler

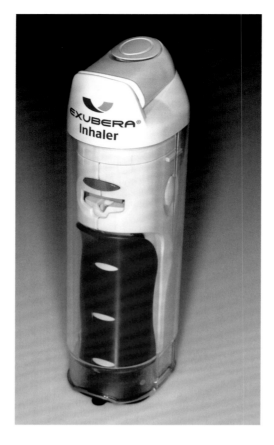

allow the insertion of genes without the host's immune system rejecting it. Continued work in unraveling the complexities of the human genome would be of great use in developing effective and safe gene therapy which promises to revolutionize medicine by its widespread use.

Although most infectious diseases have been controlled by vaccines, there are still several lethal diseases for which vaccines are not available. These include malaria and HIV-AIDS and some forms of tuberculosis and influenza. There is a lot of research going on to produce vaccines against these diseases and the 21st century holds the promise of effective means to control them. Already in 2006, a vaccine against a form of cervical cancer (caused by a kind of virus) has been developed. Some trials of vaccines against malaria are also going on, though these do not look very promising as yet. Vaccines against new diseases like avian influenza are also in the making.

The 21st century will also see continued research in pharmaceuticals and medical devices. New forms of existing medicines will be developed which will be more effective and possibly affordable. Already insulin which can be delivered by an inhaler is available, making it much easier for people with diabetes to control the disease. Similarly, medical devices will improve enormously with the use of newer and safer materials like those made with nanotechnology. Surgical interventions would also experience a change with safer and shorter surgical procedures using technologies like

improved laparoscopes and non-invasive method Finally, cures for diseases like cancer, which hav hitherto proved to be very challenging, might b found with advances in biotechnology.

In diagnostics, labs-on-chips would make quic and accurate diagnosis available at least in th developed countries. Improvements in existin diagnostic methods like X-rays, CT scans and MR would allow health professionals to understand th nature of disease better and provide more effectiv and timely treatment. Already, the introductio of functional MRI (fMRI) in medical researcl especially in the functioning of the brain, is yieldin promising results.

NANOTECHNOLOGY: THE TECHNOLOGY OF THE FUTURE

A big breakthrough in the 21st century is expecte to be in the field of nanotechnology, being th technology of manipulation and fabricatio of materials at the scale of nanometers o billionths of a meter. The last few years hav seen tremendous developments in this area, an many new techniques and technologies for us in nanotechnology are now available. Althoug the real potential of nanotechnology has yet t be realized, there are already several material which are available for the consumers. Example include synthetic fibers which are stain-repellen and wrinkle-free, or ceramic and glasses wit easy-to-clean and scratch-resistant surfaces.

However, the next few years will se nanotechnology impacting almost all spheres o human activity. In manufacturing, nanotechnolog will make possible the fabrication of lighter an much stronger materials. These materials wil revolutionize transportation with safer and mor fuel efficient vehicles. These can also be mad programmable and hence could change the fac of construction in completely unknown ways For instance, corrosion-resistant and scratch resistant materials with unique thermal propertie would make manufacturing and constructio totally different from what it is now. The use o nanotechnology to make more efficient, cos effective solar panels could revolutionize the us of renewable energy, too.

Given the size of nano-devices, it would b possible to incorporate sensors and processor into them and then fix them in many differen environments. For instance, one could thinl of nano-sensors to detect pollution, changes i temperature or even changes in levels of certair chemicals. These sensors would be exceptionall sensitive—as chemical sensors, they can detec contamination by a single molecule—and henc an be very useful. What is more, these sensors ould be enabled to transmit the data remotely y wireless to the Internet and then onward to ny location.

In medicine, nanotechnology could have a ariety of applications. Nano-sensors could detect he presence of a few cancerous cells in the body nd hence could prevent the malignancy from preading by timely intervention. Drug delivery ould be revolutionized by using nanoparticles as he drugs could be made to reach the affected part f the body by devising nanoparticles that would ttach themselves to specific cells, for instance umor cells.

Diagnostic methods using nanotechnology would e another area of development. Nanoparticles of ifferent sizes could attach themselves to different hemicals or receptors within the body and this ould be detected using methods like fluorescence r magnetic detection. Similarly, nanotechnology ould be used to massively improve the lab-on-hips which are used for the biochemical analysis f very miniscule samples.

Nanotechnology could also be used for manufacturing devices for people with certain isabilities. For instance, it could be used to make etinal implants to transmit some information to he brain to assist blind people. Or, devices could e made of biologically safe materials to assist in neurological and motor disorders. If successful and afe nano-mechanical devices like motors could be made, they could even repair or replace certain organs, leading to a revolution in medicine.

In information technology, storage of data s an area where nanotechnology could have a huge impact. Although current technologies,

namely those using the giant magnetoresistance phenomenon, have already increased data storage densities substantially, nanotechnology opens up the possibility of increasing this by several orders of magnitude. Some of these technologies are already in prototype stage and it is only a matter of time before they will appear on the market. In the field of computer memory, magnetic random access memory, a device which will use nanometer thick magnetic layers, is already being developed, and this will be a giant step toward cheaper and higher memory devices.

Nanotechnology thus promises to be an area with the potential to impact almost all spheres

of life in the 21st century. However, as with all technologies, there are risks associated with the uncontrolled use of nanotechnology, the primary concerns being their impact on health and environment. This is because there is already evidence that superfine, ultra small particles pose a considerable health risk if inhaled. Nanoparticles are particularly harmful in that sense; because of their extremely small size, there is the possibility that the human body could be affected at very small, even cellular scales. The use of nanoparticles, in paints for instance, needs to be studied since they could be carcinogenic and thus extremely harmful. In addition, the impact of nanoparticles and nanomaterials like carbon nanotubes and wires on the environment needs to be scrutinized as they could easily get into the food chain and hence have potentially disastrous consequences.

iPod Nano

In October 2001, Apple Computers introduced a portable music player iPod which became an instant hit with consumers. The iPod replaced the big and unwieldy players and had an iconic design which became very popular. Using a hard disk and proprietary software, it had a user-friendly interface. Apple introduced many versions of the iPod, including the ultrathin iPod Nano, and later started iTunes, an online music store which allowed users to buy single tracks at a cheap price.

MEMS nanotechnology

The incorporation of nanotechnology with electromechanical devices led to the development of Micro-Electro-Mechanical Systems (MEMS). MEMS devices are finding increasing use in many applications in medicine and manufacturing as well as consumer electronics. MEMS laser scanning technology for use in bar code scanning is seen here. This device is 40 times faster and less than 1/60th the size of traditional laser scanners.

Science Fiction—Today and Tomorrow

Science fiction is quite a recent genre of literature and cinema. The name science fiction was given to any form of fiction which deals with the impact of science and technology on human society by the American publisher Hugo Gernsback in the 1920s. Though it is important for science and technology to be plausible within the domain of the known knowledge, it need not be real. In fact, some of the best known science fiction is based on imagined technologies of the future.

The themes taken up by science fiction writers may not always be new. There are examples when such themes have been introduced by classical writers, such as in the second century AD, a trip to the moon is used as the plot for a satirical work. Much later, the works of Jonathan Swift, Voltaire and other visionaries of in the 18th century also used fantasy themes for writing satires on their contemporary social structure coming to terms with the tremendous changes brought about by the Industrial Revolution.

The pioneer of science fiction in a strict sense was the 19th century French author Jules Verne whose most important works are *Twenty Thousand Leagues Under the Sea* (1870) and *Around the World in Eighty Days* (1873). Verne wrote many novels which featured technologies that could only have bee imagined in those times—spaceships, submarine and even aircraft. His books were extremely popula making Verne an international figure. A hallmar of Verne's work was the extraordinary accurac with which futuristic technologies were described Verne continues to be extremely popular and i supposedly the third most translated author in th world, testifying to the cross-cultural fascination c the genre.

Toward the end of the 19th century, H.G. Well wrote three fascinating books, addressing theme which would continue to excite writers for man years. *The Time Machine*, *The Invisible Man* an *The War of The Worlds* were masterpieces dealin with space and time travel and encounters wit alien beings. Others around this time who als wrote about science fiction themes were C.S Lewis, Aldous Huxley, George Orwell and Edga Rice Burroughs.

The golden period of science fiction really cam in the 1920s when Hugo Gernsback, who use to publish magazines for radio and electrica hobbyists, realized that fiction which incorporate technology, real and imagined, would be popula He started several magazines which publishe science fiction stories on a regular basis, providin opportunities to many budding writers of the genre By the 1930s, science fiction had become extremel popular and there were fan clubs across the US an even in Europe.

As the genre gained popularity, there came abou a subtle change in the kind of science fiction tha was written. Many science fiction writers and editor were themselves trained scientists and engineers and thus put a stress on scientific accuracy Although a majority of the readers were teenagers

Jules Verne

Along with Hugo Gernsback and H.G. Wells, French author Jules Verne laid the foundation of science fiction. He introduced the concepts of air, space and underwater travel in his works before aeroplanes or submarines were invented. His popular works include *Journey to the Centre of the Earth* (1864).

2001: A Space Odyssey

Stanley Kubrick's motion picture *2001: A Space Odyssey*, based on the science fiction novel by Arthur C. Clarke, was a thought-provoking depiction of the perils of artificial intelligence. A blockbuster, the movie introduced several path-breaking special effects technologies.

The Lost World: Jurassic Park

Jurassic Park, a blockbuster movie based on the Michael Crichton novel of the same name, dealt with the theme of genetic engineering of dinosaurs. Pointing at the dangerous aspects of genetics, the movie played with the fears of the spectators.

he genre was a source of wonder and inspiration which possibly led many readers to take on careers in science and technology.

The atomic bomb in the mid 1940s and the launch of the Sputnik in the late 1950s were events which had a profound impact on science fiction. The devastating effects of a nuclear war were by now realized by scientists and a depiction of post-nuclear holocaust became a popular theme. This was also the time when writers moved away from technological futurism toward writing about the impact of science on society in the future. The cold war, which was a dominant theme in the West, was reflected in this genre too—communist paranoia and a fear of the Big Brother were popular themes.

In this period, the most popular writers of science fiction were Robert Heinlein, Isaac Asimov and Ray Bradbury—all Americans—and a Briton, Arthur C. Clarke. In the non-English world, the Polish author Stanislaw Lem and the Italian Italo Calvino, along with several Soviet writers like the Strugatsky brothers fascinated their readers.

In the 1960s the television series *Star Trek* became hugely popular. In fact, *Star Trek* acquired a cult status among science fiction fans and became a part of the popular imagination of a whole generation. In motion pictures, several movies were made during the 1960s which opened up new horizons in the depiction of the future, *2001: A Space Odyssey* and *Fahrenheit 451* being the most popular ones. The success of these movies and the television series led to Hollywood taking up the genre in a big way. *Close Encounters of the Third Kind*, *Star Wars* and *E.T.* were among the biggest blockbusters of the late 1970s and 1980s, spawning a huge industry in sequels as well as merchandizing of products.

The tremendous growth in the field of computers during the 1970s and 1980s was also reflected in the themes which science fiction addressed. If the works of the 1950s and 1960s had been predominantly focused on aliens and space travel, the 1980s saw the emergence of cyberpunks and artificial intelligence in literature and motion pictures. *Blade Runner*, a hugely successful movie made in 1982, is an example of this sub-genre which continues to attract audiences.

Computer technology also had a major impact on the production of science fiction movies and television shows. Toward the end of the 20th century, the special effects in movies had become very sophisticated. The technology for special effects allowed a very realistic portrayal of unreal characters and machines, and this added to the attraction of this movie genre. *Jurassic Park*, an enormously successful movie based on the reconstruction of dinosaurs using recombinant DNA technology, was a milestone in this direction. *Terminator*, *Matrix* and *Transformers* were all movies in which realistic special effects played an important role.

The 21st century society is already a highly technological society. Space travel, satellites, nuclear power, computers and even robots are now part of everyday life. There is thus a feeling among some that the charm of science fiction would be lost in the current scenario. However, the pace of scientific and technological innovation in the 21st century is certainly going to be much faster than before. With new areas and dimensions opening up in almost all areas of science and technology, there will always be enough fundamental science for authors of science fiction to imagine what the future might hold.

LABORATORIES AND THINK TANKS

For the large part of human history, technological innovation has come from individuals tinkering or experimenting by themselves. These could be craftsmen, farmers, hunters, priests, courtiers or even gentlemen of leisure. With the growth of universities in the West after the 11th century AD, these new institutions gradually became the center for learning and new ideas. Nevertheless, it was only in the 20th century that technological innovation got institutionalized.

Thomas Edison established the first industrial laboratory at Menlo Park toward the end of the 19th century. The idea was to have many engineers and technicians interacting and developing new ideas and techniques. The basic infrastructure and equipment which individuals might find hard to procure were made available at the laboratory. This was a stupendously successful model, and within a few years, the laboratory gave the world the phonograph, the incandescent bulb and several other inventions.

This idea was taken up by AT&T which established a research laboratory named the Bell Telephone Laboratories in 1925. Bell Labs became one of the most successful examples of a research laboratory. Spread over many locations in the US, Bell Labs was responsible for thousands of inventions, including several revolutionary ones like the transistor and the photovoltaic cell. Scientists at the lab also carried out work in fundamental science and discovered electron diffraction, radio astronomy, information theory and fractional Hall effect. In fact, so far six Nobel Prizes have been awarded to work done at the Bell Labs.

There are several other examples of very successful research laboratories sponsored by industry. These include the IBM research labs and the Xerox Palo Alto Research Center (PARC). Established in 2002, PARC pioneered technologies such as laser printing, Ethernet, the graphical user interface (GUI) and ubiquitous computing. The basic idea behind these institutions was to give scientists and engineers the freedom and opportunity to pursue their ideas without undue commercial pressures. Of course, the labs made huge amounts of money by licensing their inventions but this did not mean that only commercially viable ideas were encouraged.

Apart from the private sector, most countries also have research laboratories which are set up and funded by the government. In the US, there is a network of National Laboratories and Institutes which carry out basic and applied research in almost all fields. In most cases, these institutions are set up near research universities to gain from the presence of qualified human resources. Several of the laboratories carry out classified research with defense applications but there are many which do fundamental research.

Some universities have also founded research institutes to work as independent non-profit entities and carry out research on contract. A prime example of this is the SRI International, which was founded by Stanford University in 1946 and has developed several key inventions in the areas of communications, networks and computers.

Science and technology in the 21st century is an enormously huge and complex body of knowledge. The very vastness of any specific field or subject in science and technology leads to a growth in specialization. It is no longer possible for any individual to master the complexities of many subjects as it has been the case till the Industrial Revolution. Furthermore, the tools required for technological innovation and experimentation are such that they are beyond the reach of individuals. Hence state or corporate sponsored institutions are where almost all of technological development takes place. As the world of science and technology grows more and more complex, there would probably be a move toward more specialized laboratories rather than general research establishments.

Bell Labs

Bell Telephone Labs or Bell Labs was one of the most successful examples of an industrial research laboratory. Although initially set up to carry out applied research into telephony related devices, in its later years it promoted basic as well as applied research in many fields. Many researchers working at the Bell Labs went on to win the Nobel Prize. A researcher is seen here working on a communicative oscilloscope at the Bell Labs.

NASA

NASA, established in 1958 by the US government, is an example of a huge scientific enterprise which has been responsible for developing many technologies in space science. Among its crowning achievements is the development of the reusable space vehicle or the space shuttle which, although it proved to be much more expensive than the projected costs, has been enormously useful for space research.

A FUTURISTIC WORLD?

Human beings have come a long way since the first Homo sapiens hunted for food in Africa millions of years ago. Technology, the purpose-driven use of natural and artificial materials and objects, has progressed immensely since Paleolithic times. Although it is tempting to see the tremendous technological achievements of mankind in the last century as unique, it is important to keep a historical perspective.

As seen in this book, the last 12,000 years experienced a series of revolutionary innovations to improve life. The development of an effective tool to kill animals during a hunt must have seemed as revolutionary to the Neolithic hunter as smart bombs appear now. The domestication of cereal plants in the Middle East was probably an event of much larger significance than the development of GM foods, since it changed the course of human history. Similarly, the advent

technological innovation is now way beyond the reach of individuals. This trend has been specially pronounced in the latter part of the 20th century when scientific and technological research became institutionalized.

It is expected that the 21st century will see further changes in the way technological innovation takes place. The advent of the Internet and fast and cheap communications have already enabled a geographical dispersal of research, a trend which will grow in the coming years. Dispersal of technological innovation globally would also become faster and access to advanced technologies, especially in consumer goods, would be easier and more widespread.

It is an exercise fraught with dangers to predict what kinds of technologies would emerge in the 21st century. The creativity of the human mind is essentially unlimited and thus technological prediction is more often than not wrong. What shape technology will take and how it will impact

of writing in Mesopotamia could be taken as an event of enormous significance, possibly even more important in the long term than the invention of the computer in modern times. One could think of many such examples throughout history where key inventions and innovations have changed the course of human history.

However, the uniqueness of recent times, especially since the Industrial Revolution, lies in the speed with which technological innovation has disseminated through society. In the 20th century, the organization of technological innovation underwent a profound change. Research and development were no longer carried out by isolated individuals but by large research and development establishments. These could be industrial laboratories, universities or government laboratories. The infrastructure needed for

human society cannot be predicted with any degree of certainty. After all, even the most brilliant scientists in the early 19th century could not have predicted space travel and the microprocessor. Nevertheless, it can be said that technological progress would go on at an even faster pace in the coming years and possibly make life easier for the vast majority of human beings, with revolutionary developments in medicine, transportation and communication technologies.

However, with nature beginning to protest, as evident in global warming and the depleting ozone layer, it is important to bring down the pace of technological advances for the sake of future generations. As the work in developing alternate technologies to save the environment has already begun in the 21st century, it needs to be made more popular in the coming years.

What does the future hold?

Looking at the pace of technology advancement in the 21st century, it seems that the future will bring about great leaps in medicine, communications and transportation. However, environmental issues will now take a front seat in all new developments.

GLOSSARY
INVENTIONS AND INVENTORS

Adding Machine: In 1888, the American bank clerk William Burroughs (1857–1898) received a patent for an adding machine which reduced the drudgery of repeated calculations. Burroughs later formed a company that went on to produce electric calculators and later computers.

Air Conditioner: The first electrical air conditioner was invented by the American engineer Willis Carrier (1876–1950) in 1902. The machine controlled the temperature and the humidity and was initially used in factories where temperature and humidity control was crucial. Subsequently, the use of air conditioners became very widespread in domestic and commercial establishments.

Airplane: The Wright brothers, Orville (1871–9148) and Wilbur (1867–1912), demonstrated the first powered, heavier-than-air flying machine in 1903.

Arc Lamp: In the first decade of the 19th century, the famous English scientist Sir Humphry Davy (1778–1829) demonstrated the first arc lamp by using carbon pieces, separated by air and connected to a voltaic pile or battery. The arc lamp principle was used later with different gases to get different colored light.

Artificial Satellite: The first satellite Sputnik I was launched in 1957 by the Soviet Union, thereby opening up space for human beings.

Astrolabe: Hipparchus (190–120 BC) invented the astrolabe which was of great use for doing calculations in astronomy and also in navigation. The astrolabe continued to be in use for many centuries with modifications.

Atomic Clock: The first atomic clock, a device which uses the transitions of atoms to accurately determine intervals of time, was made at the US National Bureau of Standards in 1949. Atomic clocks are now used as time standards and are extremely accurate.

Bakelite: The American chemist Leo Baekeland (1863–1944) synthesized the first plastic from completely synthetic components in 1908. Bakelite, as the material was called, had insulating properties and was used in electrical components. It was also used for making toys, kitchenware and even jewelry.

Ballpoint Pen: In 1938, a Hungarian journalist, László Bíró (1899–1985), patented the first ballpoint pen which used a roller mechanism on the tip to get ink from a cartridge. Ballpoint pen replaced fountain pens to a large extent over the next few decades since they were convenient and there were no ink spills.

Bessemer Process: Henry Bessemer (1813–1898), an English inventor, patented a process of making steel much faster and inexpensively than before. Bessemer process allowed a widespread use of steel for a variety of purposes like construction, weaponry, machine building and cutlery.

Braille: The French inventor Louis Braille (1809–1852) invented the Braille system in 1821. The system is widely used by the visually impaired for reading and writing.

Bronze: Around 3000 BC, in Mesopotamia, tinstone and copper were reduced together to produce bronze which was more useful than pure copper. It was much harder than copper and could be used for weapons and tools.

Buoyancy: Archimedes of Syracuse (287–212 BC) discovered the principle of floatation and the laws of buoyancy.

Canals: Around 4000 BC, Mesopotamians built the first canals to use the water of Tigris and Euphrates to irrigate the fields.

Cannons: In the 11th century, the Chinese used bamboo tubes—which were later replaced by cast iron tubes—to throw lead pellets. This principle was used in throwing heavy metal filled with gunpowder, or cannons.

Carburetor: Carburetor, a simple device which mixes air, with petrol was invented by the Hungarian engineer Donát Bánki (1859–1922) in 1893. The carburetor helped the development of the automobiles immensely since it provided an easy method of creating a fuel–air mixture which was needed for combustion.

Chariot: Around 2000 BC, when the Mesopotamians had contact with their northern neighbors from the central Asian steppes who had domesticated the horse, they developed the horse-drawn chariot.

Circumference of the Earth: Eratosthenes (276–194 BC), a Greek mathematician, discovered a method of finding the circumference of the earth by using the elevation of the sun at two different places. His measurement of circumference was 84 per cent correct.

Cochlear Implant: The American physician William House (1923–2012) made the first cochlear implant in 1961. This electronic device is implanted inside the ear to improve the hearing of the hearing impaired.

Colt Revolver: In 1835, the American inventor Samuel Colt (1814–1862) patented a kind of revolver which used an ingenious mechanism to rotate the cylinders. Colt's revolver was immensely popular in the American West since it was the first practical repeating firearm.

Combine Harvester: A machine which combines the agricultural tasks of harvesting and threshing of corps was invented by the American inventor Hiram Moore (1798–1858) in 1834. The combine harvester was extremely useful, especially in the large farms of North America where labor was in short supply. The earliest combines used animal power which was later replaced by steam and finally by the internal combustion engine.

Compact Disc: Invented by James Russell (1931–) in 1965, the compact disc became popular when it was released in 1982 Phillips Consumer Electronics. The compact disc allowed for a much higher fidelity of reproduction and soon replaced magnetic tapes in the music industry.

Computer: The first general, all-purpose computer ENIAC was built for the US Army in 1946. The ENIAC used vacuum tubes and could be programmed.

Computerized Automated Tomography (CAT): CAT scanning, a technique which uses multiple X-ray images to produce an accurate image of the human body, was invented by Godfrey Hounsfield (1919–2004), an English engineer, 1972. Since then, CAT scanning has become an importa diagnostic tool in medicine.

Computer Mouse: The American engineer Dougl Engelbart (1925–) made the first computer roller ba mouse in 1963. The mouse became an essential device wit computers after the advent of the Graphical User Interfac (GUI) in the 1980s.

Copper: Around 3500 BC, somewhere in the Middle Eas copper was reduced from its ores and melted in stor crucibles to be cast into molds.

Cotton Gin: A simple machine to separate out cotton fibe from the seedpods was invented in 1793 by the America inventor Eli Whitney (1765–1825). The cotton gin increase the production of cotton and was important for establishin cotton plantations in the American South.

Crankshaft: The Islamic polymath Ibn Ismail ibn al-Razza al-Jazari (1136–1206) invented the first crankshaft in th late 12th century AD. The crankshaft proved to be of grea importance during the steam age.

Crossbow: Chinese sources from the fourth century BC ta about the use of a crossbow catapult. The crossbow was remarkably effective weapon for its time.

Diesel Engine: The German automotive engineer Rudo Diesel (1858–1913) in 1893 invented a kind of interna combustion engine which caused combustion by compressio of the air. Diesel engines are more efficient and give greate power than gasoline engines.

Digital camera: Steven Sasson (1950–) created the firs digital camera in 1975.

DNA Sequencing: In 1975, Frederick Sanger (1918–), a English biochemist, discovered a way to sequence the double helix molecule of DNA. The sequencing or determining th order of the bases in the molecule proved to be importar in understanding the functioning of the DNA.

DNA Structure: In 1953, two Cambridge scientists James Watson (1928–) and Francis Crick (1916–2004 deciphered the structure of DNA, the genetic material i all living organisms. The understanding of the structure of DNA could be seen as the beginning of a revolution i genetics and biology.

Dynamite: In 1866, the Swedish chemist Alfred Nobe (1833–1896) discovered a way of making nitroglycerine a highly explosive substance, safe. Dynamite, a mixture o nitroglycerine and other chemicals, is used extensively no only in warfare but also in mining and construction.

Electric Battery: The Italian scientist Alessandro Volta (1745–1827) invented the electric storage battery in 1800 The battery used chemical interaction to produce an electric voltage between its terminals. The unit for voltage is the Volt in Volta's honor.

Electric Motor: The first electric motor was made by the English scientist Michael Faraday (1791–1867) in 1821 The electric motor, a device which used magnetism to conver ectric energy into mechanical energy, was later improved on and became the standard provider of power.

ectric Dynamo: In 1831, Michael Faraday (1791–1867) nstructed the dynamo, a device for generating electric rrent by using electromagnetism. The dynamo made it ssible to generate electricity by using mechanical power.

ectrical Telegraph: The first commercial electrical egraph was patented by two English scientists, William oke (1806–1879) and Charles Wheatstone (1802–1875) 1837. The telegraph revolutionized communication, pecially for businesses, as it could be used to get formation across hundreds of miles almost instantly.

ectrocardiograph (ECG/EKG): The Dutch physiologist illem Einthoven (1860–1927) invented the first practical ectrocardiograph in 1903. This machine, which records e electrical activity of the heart, has proved to be very eful in diagnosing the ailments of the heart.

ectroencephalograph (EEG): The German physiologist ans Berger (1873–1941) invented the EEG, a machine hich could record the electrical activity of the brain. The EG is used not only for diagnosis but also as a research ol to understand the working of the brain.

evator, Safety Brake: In 1853, the American inventor isha Otis (1811–61) invented an elevator brake which evented the elevator car from falling in case the cables oke. This made the elevators much safer, which were bsequently used extensively in high-rise skyscrapers.

thernet: In 1973, Robert Metcalfe (1946–) and David oggs invented the Ethernet, a computer networking chnology which became the dominant mode for networking computers in the 1980s and later.

rmented drinks: Mesopotamian seals show the first presentation of fermenting date palm and barley to make rmented alcoholic drinks around 2500 bc.

untain Pen: The first capillary-fed fountain pen was ade by the American insurance broker Lewis Waterman 837–1901) in 1884. These pens were much easier to use d rarely leaked while writing.

riction Match: In 1827, the English chemist John Walker 781–1859) discovered that a mixture of antimony lphide, potassium chlorate, starch and gum could be sed to make a friction match which would ignite on being bbed against a rough surface.

llerenes: In 1985, three scientists, R. Curl (1933–), R. malley (1943–2005) and H. Kroto (1939–), discovered new form of carbon called fullerenes. Fullerenes opened a whole new world of carbon nanotubes and carbon ckyballs with unique properties.

usion Bomb: In 1952, the Hungarian–American physicist dward Teller (1908–2003) and the Polish mathematician tanislaw Ulam (1909–1984) designed the first successful sion bomb or hydrogen bomb.

eiger Counter: In 1908, the German scientist Hans Geiger 882–1945), together with the New Zealand physicist rnest Rutherford (1871–1937), developed a counter to etect radioactivity. The Geiger counter is used extensively ll today and has proved to be a very useful device in the uclear industry.

eometry: Around 300 bc, Euclid of Alexandria set down is postulates of geometry and wrote *Elements*, a collection f theorems. *Elements* is considered one of the most fluential books in mathematics.

Global Positioning System (GPS): The US Department of Defense in 1995 launched of a global, satellite based navigation system, the Global Positioning System. The GPS allowed an accurate determination of position anywhere on the earth and is now finding use in the civilian domain.

Gregorian Calendar: On February 24, 1582, Pope Gregory XIII (1502–85) issued a papal bull to introduce a new reformed calendar to take care of the discrepancy between the seasons and the calendar dates. This was done by dropping some days and having rules for leap years.

Gunpowder: One of the "four great inventions" of ancient China, gunpowder was invented in the ninth century. Gunpowder made its way from China to Persia and then to Europe over the next few centuries.

Handguns: The first handguns were used in China in 1288 ad.

Helicopter: The helicopter, a vertical take-off and landing machine powered by rotors, was invented by the Russian–American aviation engineer Igor Sikorsky (1889–1972) in 1939. Although several designs had been tried out before, Sikorsky's design was the first successful and practical one. Helicopters were used extensively in warfare and also for civilian transport after World War II.

Holography: In 1947, the Hungarian physicist Dennis Gabor (1900–79) invented a process to produce a three-dimensional image of objects. Holograms could not be produced easily until the invention of laser in the 1960s.

Hovercraft: Christopher Cockerel (1910–99), an English engineer, first developed the principle of transporting on air cushion in 1952. This led to the invention of hovercraft.

Hot-air Balloon: The first machine to enable human beings to be airborne was the hot-air balloon invented by the French brothers Joseph (1740–1810) and Jacques (1745–99) Montgolfier in 1783.

Incandescent Bulb: In 1880, Thomas Alva Edison (1847–1931) demonstrated that using a carbon filament in a light bulb would give it a much longer life. Prior to this, the filaments of electric bulbs were prone to burn-outs and thus not very practical.

Insulin: In 1920, the Canadian doctor Frederick Banting (1891–1941) discovered insulin (derived from pig's pancreas) which plays a crucial role in the metabolism of sugar in the human body. Banting's discovery was responsible for the use of insulin in treating diabetes.

Integrated Circuit: In 1959, the American engineer Jack Kilby (1923–2005) was granted a patent for developing a complete electronic circuit on a germanium wafer. The integrated circuit, as this invention was called, was responsible for the growth of miniature electronics and is extensively used today.

Internal Combustion Engine: The German Karl Benz (1844–1929) built the first automobile powered by a four-stroke gasoline engine in 1885. Benz and other engineers like Daimler and Maybach later introduced many innovations in the technology of the internal combustion engine.

Internet: Robert Kahn (1938–) and Vinton Cerf (1943–) invented the TCP/IP protocol in 1983. This protocol, or set of rules, is used to transmit information over the Internet.

Iron: Sometime between 2000 bc and 1500 bc, the Hittites in Eastern Turkey discovered the use of iron. Over the next few centuries, this technology spread across Mesopotamia and Egypt.

Jet Engine: The British air force engineer Frank Whittle (1907–96) patented the first jet engine in 1932. The first jet engine was made by Whittle in 1937 but it was many years before it replaced the propeller as the power source in aircraft.

Kinetoscope and Kinetograph: In 1893, Thomas Alva Edison (1847–1931) demonstrated two pioneering inventions to shoot and view motion pictures. These inventions, which underwent many improvements in the next few decades, led to the birth of the movie industry.

LASER: The American physicist Theodore Maiman (1927–2007) produced the first working laser (Light Amplification by Stimulated Emission of Radiation) in 1960. Lasers find extensive use in a variety of fields like communication, surveying and consumer electronics.

Light Emitting Diode (LED): In 1962, the American scientist Nick Holonyak Jr. developed the first light emitting diode. LEDs are used in communication, display screens and also electronic circuits.

Lightning Conductor: The American statesman and scientist Benjamin Franklin (1706–1790) invented the lightning rod in 1752. Franklin had experimented with atmospheric electricity before and had shown that lightning was basically an electric spark.

Liquid Crystal Display (LCD): In 1970, the Swiss company Hoffman-LaRoche patented the use of the nematic effect in liquid crystals and started licensing the technology to produce liquid crystal displays (LCDs). LCDs became the standard display in many consumer electronics like wrist-watches and game consoles.

Liquid Fuel Rocket: The American space scientist Robert H. Goddard (1882–1945) in 1913 obtained a patent for a rocket which used liquid fuel—a mixture of gasoline and liquid nitrous oxide. The liquid fuel rocket was an important milestone in rocket science.

Lithography: A technique to print text or art onto paper, lithography was invented in 1796 by the German author Alois Senefelder (1771–1834). Lithography is essentially a chemical technique that has been refined for use in the manufacture of chips and other devices.

Locomotive, steam: In 1801, the English mining engineer Richard Trevithick (1771–1833) made the first working locomotive which ran on steam power. The engine used a crankshaft to convert linear motion of the piston into rotary motion. The design of the locomotive underwent many changes in the 19th century.

Logarithm: John Napier (1550–1617), an English mathematician, invented logarithms in the first decade of the 17th century. Logarithms were extremely useful in carrying out complicated calculations and continued to be used extensively till the advent of calculators and computers.

Loom: Egyptians and Mesopotamians used a simple form of a loom to weave a kind of coarse cotton cloth around 3000 bc.

Magnetic Compass: Shen Kuo (1031–1095 ad), a Chinese engineer in the 11th century, used a magnetized needle floating in water to make the first magnetic compass. The compass revolutionized navigation.

Magnetic Resonance Imaging (MRI): In 1971, the American scientist Raymond Damadian (1936–) developed the first magnetic resonance imaging device which could be used to give detailed images of the human body.

Marine Chronometer: In 1737, the British clockmaker John Harrison (1693–1776) invented an accurate marine chronometer. This device allowed the mariners to accurately determine their longitude on the high seas and thus was instrumental in long-distance sea exploration in the 18th century.

MASER: MASER (Microwave Amplification by Stimulated Emission of Radiation) was invented by the American physicist Charles Townes (1915–) in 1953. The MASER produced a remarkably coherent beam and led to the development of LASERs subsequently.

Mercury Thermometer: In 1724, the German engineer Daniel Fahrenheit (1686–1736) first used mercury in a thermometer. He was also the first person to devise a standardized scale for use on thermometers.

Microphone: In 1876, Thomas Alva Edison (1847–1931) invented the carbon microphone. Although the microphone had been invented earlier, carbon microphones became the standard in communication for several years since they were low cost and gave a high output.

Microprocessor: In 1971, the American company Texas Instruments patented the microprocessor, a programmable electronic component which uses digital electronics and forms the heart of modern computers.

Microscope: The microscope was invented In the first decade of the 17th century. It is not clear who was the first person to use two lenses in a particular combination to magnify extremely small things, but the credit is usually given to Hans Lippershey (1570–1619).

Microwave Oven: The American engineer Percy Spenser (1894–1970) accidentally discovered that microwaves could be used for cooking in 1945. He patented the microwave oven in 1946 but it was not until the 1980s that microwave ovens became cheap and portable enough to be popular.

Miners' Safety Lamp: Sir Humphry Davy (1778–1829) in 1815 introduced a modification in the lamp used by miners in the coal mines. The older lamp was responsible for a large number of explosions in the mines as the combustible gases in the mines caught fire due to the lamp. The new design by Davy made the lamp much safer and was responsible for saving many lives.

Morse Code: Samuel Morse (1791–1872), an American painter, pioneered the single wire telegraph and also devised a code for use in telegraphy in 1836. This code became the standard for communication and was used across the world.

Mosaic Web Browser: In 1993, Marc Andreessen (1971–) developed the first browser called Mosaic. The Mosaic allowed the user to use the World Wide Web easily and was responsible for its tremendous growth.

Nuclear Pile: In 1942, the first controlled nuclear chain reaction was achieved by the Italian–American physicist Enrico Fermi (1901–1954) as part of the Manhattan Project to make a nuclear weapon. The controlled chain reaction is an important part of a nuclear reactor.

Nuclear Weapons: The Manhattan Project was successful in producing two nuclear weapons in 1945. These weapons, which were based on nuclear fission, were dropped on Japan and proved to be the deadliest weapons used by humans in history.

Oral Contraceptive Pill: The chemists Carl Djerassi (1923–), Luis Miramontes (1925–2004) and George Rosenkranz (1916–), working in Mexico, developed the first oral contraceptive pill by synthesis in 1951. However, it was several years before this revolutionary medicine was approved for general use and became highly popular, especially in the US.

Packet Switching: The American engineer Paul Baran (1926–2011) in 1962 proposed the technique of packet switching. This technique, in which packets of data move between nodes in a network, is used in the Internet.

Paper: Paper, made from wood pulp, was first introduced in China around the 1st century ad. Prior to this, papyrus and parchment had been used for writing. Paper is considered one of the "four great inventions" of ancient China.

Pasteurization: The process of sterilizing liquids by heating it was first discovered by the French chemist Louis Pasteur (1822–95) in 1862. The heat treatment does not kill all the bacteria but reduces their numbers significantly without altering the taste of food. It is used extensively till date, especially with milk.

Penicillin: The English microbiologist Sir Alexander Fleming (1881–1955) in 1928 accidentally discovered that a particular kind of fungus had the property to kill bacteria. The discovery of antibiotics was responsible for saving millions of lives especially during World War II.

Pendulum Clock: In 1656, the Dutch scientist Christiaan Huygens (1629–95) invented the pendulum clock. The pendulum clock remained the most accurate way of determining time till well into the 20th century.

Personal Computer: In 1972, Xerox PARC developed the first personal computer, the Xerox Alto. This particular invention was not available commercially. The IBM PC was the first widely available personal computer.

Phonograph: In 1877, the American inventor Thomas Alva Edison (1847–1931) demonstrated a device to record and playback sound. The phonograph used tinfoil cylinders initially but these were later replaced by flat records.

Photography: In 1826, the French inventor Nicéphore Niépce (1765–1863) made the first permanent photograph using chemical means. His co-worker L. Daguerre improved the process to invent the daugerreotype in 1839.

Pneumatic Tire: The Scottish veterinarian John Boyd Dunlop (1840–1921) developed the first pneumatic or air filled tire in 1888. The pneumatic tire proved to be of great use in the future development of automobiles.

Polaroid Camera: Edwin Land (1909–1994), an American scientist, invented the camera with a self-developing film in 1947. The camera, called the Polaroid camera after Land's company which made them, became very popular since it produced instant photographs.

Polio Vaccine: Jonas Salk (1914–95), an American biologist, developed the first vaccine against polio. The polio vaccine played a vital role in eradicating polio from the West and also reducing its incidence among children around the world.

Polymerase Chain Reaction (PCR): In 1983, Kary Mullis (1944–), an American biochemist, invented the Polymerase Chain Reaction. This technique allowed the amplification of a small piece of DNA and hence made the analysis and manipulation of DNA possible. It is widely used in biotechnology now.

Power Loom: Edmund Cartwright (1743–1823), an English clergyman, invented the power loom in 1784. The power loom was a key invention of the Industrial Revolution and initially used water as the source of power. With the advent of steam engines, steam powered looms became the mainstay of the textile industry in England.

Plastic Surgery: Susruta, an Indian surgeon, practised plastic surgery sometime in the sixth century bc. He also wrote a book on surgery called *Susruta Samhita* which detailed many surgical procedures.

Plow: The share and the sole were amalgamated into a single piece to be pulled by yoked animals in Mesopotamia around 3000 bc.

Pressure Cooker: Denis Papin (1647–1712), a French engineer, invented the steam digester, a primitive form of the pressure cooker in 1679. The steam digester was used initially to extract fat from the bones by using high pressure steam. The digester design was later instrumental in the design of the steam engine.

Printing Press and Movable Type: Johannes Gutenberg (1400–68 ad), a German goldsmith, invented the first movable type and a printing press in the 15th century. The invention of the printing press had revolutionary implications and played an important part in the rise of Protestantism.

Printing, Woodblock: The Chinese invented the woodblock printing in the ninth century ad and this was used to reproduce the teachings of Buddha in the *Diamond Sutra*. By the 10th century, the first ceramic movable type was also introduced in China.

Radio: In 1893, building on the work of Heinrich Hertz, the Serbian inventor Nikola Tesla (1856–1943) demonstrated the first wireless transmission or radio communication. Although others were working on the development of the radio, Tesla incorporated all the elements of a radio system into his device.

Radio Telescope: The radio telescope was invented by the Karl Jansky (1905–50), an American radio engineer, in 1930. The radio telescope proved to an invaluable tool for astronomy subsequently.

Recombinant DNA: Paul Berg (1926–), an American biochemist, first created recombinant DNA by splicing and combining two pieces of DNA in 1972. Recombinant DNA techniques opened the way for biotechnology.

Saddle: The earliest forms of saddle made their appearance around 800 bc in the central Asian region. The saddle made the use of the horse for human transport easier.

Safety Pin: Although a form of safety pin had been used since antiquity in many cultures, the American mechanic Walter Hunt (1796–1859) in 1849 reinvented the design of using a steel wire to produce a pin which could be used to attach two pieces of fabric together.

Scanning Tunneling Microscope: In 1981, G. Binnig (1947–) and H. Rohrer (1933–) invented the scanning tunneling microscope. This instrument is able to produce images at the submicroscopic level and is extremely useful in understanding the structure of materials and also in nanotechnology.

Seal: Around 4000 bc, somewhere in Mesopotamia, geometric impressions were cut in stone or fired clay to create a seal. The first seals served as identification of the owner of goods.

Seed Drill: Although primitive forms of seed drills had been in use for some time, it was the English agriculturist ethro Tull (1674–1741) who invented the modern seed drill in 1701. The seed drill was an important development in agriculture during the Industrial Revolution since it increased the yield and also reduced the labor required for sowing.

Sewage systems: Around 2600 BC, the planned cities of Mohenjodaro and Harappa in the Indus valley develop artificial sewage systems, to carry the waste from the residential areas.

Sewing Machine: Elias Howe (1819–67), an American inventor, was granted the first patent for a sewing machine in 1844. The sewing machine was an important invention since it speeded up sewing and also subsequently allowed it to be carried out in homes.

Sextant: The sextant, an instrument to measure the elevation of heavenly objects in the sky, was invented by the English mathematician John Hadley (1682–1744) in 1730. The sextant proved to be an invaluable aid for navigators since it allowed them to accurately determine the latitude.

Shrapnel Shell: In 1784, Henry Shrapnel (1761–1842), an English artillery officer, developed an anti-personal weapon—it was a hollow cast iron sphere filled with a mixture of iron balls and powder together with a fuse. The fuse allowed the sphere to explode near the target and the explosion sent the iron balls off at a high speed, increasing the potency of the weapon.

Silk: Around 3000 BC, the first evidence of woven silk cloth is found in China. Silk manufacture remained a monopoly of the Chinese for many centuries.

Slide Rule: William Oughtred (1575–1660), a Cambridge mathematician, invented the first slide rule in 1630. The slide rule, a device which allowed arithmetic calculations to be done quickly, became an indispensable tool for engineers and scientists.

Solid-fueled rockets: In the ninth century, Chinese chemists discovered that a combination of chemicals gave a black powder which could be used to propel arrows. Thus earliest solid-fueled rockets were thereafter used in warfare between the Chinese and the Mongols.

Sonar: The first patent for underwater echo ranging was awarded to the English meteorologist Lewis Richardson (1881–1953) in 1912. The sonar, as this device has come to be known, is of great use for navigation and location of submarines.

Spectacles: Although corrective lenses had been used in Iran in the ninth century, the Italian Salvino D'Armate made the first wearable eye glasses in 1284 AD.

Spinning Jenny: In 1764, James Hargreaves (1720–78), an English carpenter, invented the spinning jenny. This simple machine increased the productivity of workers in the textile industry manifold since it allowed workers to work on many spools at the same time.

Spreadsheet: Dan Bricklin (1951–) created the first spreadsheet program VisiCalc in 1978. The spreadsheet program was responsible for the spread of computer use in businesses.

Stainless Steel: In 1913, the English steel-maker Harry Brearley (1871–1948) invented the process to make rust-less or stainless steel. The process involved adding chromium to the steel to make it rust-resistant. Stainless steel was used for weaponry, cutlery and machinery because of its unique properties.

Steam Engine: Thomas Newcomen (1663–1729), an English ironsmith, invented the practical steam engine in 1712. The engine was used primarily for removing water from the coal and tin mines and was hugely successful. James Watt (1736–1819), a Scottish instrument maker, made a major improvement to the Newcomen steam engine in 1769 to increase its efficiency.

Steam-powered Aeolipile: Hero of Alexandria (10–70 AD) was the first person to make a steam-powered turbine as well as weapons using air under pressure.

Steel Plow: In 1837, the American blacksmith John Deere (1804–1886) invented the steel plow. This implement was much harder than the iron plows being used and hence was able to be used on soils which were too hard for the iron plows. The steel plow was important in increasing the cultivable area available in the US.

Stethoscope: The French physician René Laennec (1781–1826) in 1816 invented the stethoscope, a simple acoustic device to listen to internal sounds of the human body. The stethoscope replaced the hearing trumpet and became an essential device for medical practitioners.

Stirrup: The stirrup was invented in the fourth century AD in China, though there were many similar contraptions in use prior to that. The stirrup was of fundamental importance for warfare since it freed both the hands of the rider for combat.

Suspension Bridges: The vast Inca Empire in the Andean region of South America had a very good system of communication. They were the first to use woven ropes and textiles to build suspension bridges to span ravines in the Andean region in the 14th century.

Talkies: In 1926, Warner Brothers, a big movie studio, introduced the Vitaphone system in motion pictures. This system allowed adding recorded sound effects and music onto the film, and introduced sound in motion pictures.

Telephone: Arguably the most valuable patent in history was granted to the Scottish American inventor Alexander Graham Bell (1847–1922) for telephone in 1876. The telephone went on to become the most important communication device in the next century.

Telescope: In 1608, the Dutch spectacle maker Hans Lippershey (1570–1619) made the first telescope by using two lenses in a tube. Lippershey's design was improved upon by the Italian scientist Galilei, who was the first person to use the telescope to view the heavens.

Television: In 1925, the Scottish inventor John Logie Baird (1888–1946) demonstrated the first working television system whereby he could transmit visuals across a distance. Television revolutionized the entertainment industry over the next few decades.

Threshing Machine: In 1784, Andrew Meikle (1719–1811), a Scottish engineer, invented the threshing machine. The threshing machine replaced human power to remove the grain from the stalks and husk, and thus increased the productivity in agriculture substantially.

Tractor: The first practical gasoline powered tractor was invented by the American inventor John Froelich (1849–1933) in 1892. Prior to this, there were steam powered machines which were not very useful. The tractor was a versatile agricultural machine which continues to be used extensively in farming.

Transistor: In 1947, three scientists at the Bell Labs, William Shockley (1910–89), John Bardeen (1908–91) and Walter Brattain (1902–87) invented the transistor which revolutionized electronics.

Triode: Lee de Forest (1873–1961), an American inventor, modified the diode by placing a grid and invented the triode in 1907. The triode, which could be used to amplify electric signals, was of great use in radio technology.

Tungsten Filament Bulb: In 1906, Franjo Hannaman (1878–1941), a Croatian, invented the tungsten filament for incandescent bulbs. The tungsten filament was long-lasting and did not lead to blackening of the glass bulb.

Typewriter: Christopher Sholes (1819–90), an American publisher, invented the first practical typewriter in 1867. Sholes also invented the QWERTY keyboard which continues to be in use even today.

Vaccination: In 1796, the English physician Edward Jenner (1749–1823) used a cowpox infected material to inoculate against smallpox, thereby inventing vaccination.

Vacuum Diode: In 1904, the English engineer John Fleming (1849–1945) invented the vacuum tube or diode. This device allowed current to flow in one direction only and was the key for development of electronic devices.

Vacuum Pump: Otto von Guericke (1602–86), a German scientist and engineer, invented the vacuum pump in 1650. The vacuum pump was subsequently improved and played an important part in the Industrial Revolution.

Videocassette Recorder (VCR): In 1956, Ampex Inc introduced the first videocassette recorder which allowed capturing motion on a magnetic tape.

Vulcanization of Rubber: The American inventor Charles Goodyear (1800–60) invented the process of vulcanization of rubber in 1839. The process of adding sulfur to rubber at high heat made it more durable, resistant to chemicals and harder.

Watch, pocket: Peter Henlein (1480–1542), a German locksmith, invented the first pocket watch in 1504.

Writing: Around 3500 BC, the Sumerians first drew pictograms on wet clay, which was later fired or dried to produce a permanent record. The origin of writing was possibly in the need to keep accounts of goods.

Wheel: Around 3500 BC, the first representations of a wheel appear in Mesopotamian pottery. The earliest wheel was made from wood and had three pieces.

Wheelbarrow: First century AD saw the appearance of the single wheeled wheelbarrow for moving material, a very useful device in construction.

Windmills: Windmills were first invented in the seventh century AD in Persia for milling grain or lifting water.

World Wide Web: Tim Berners-Lee (1955–), a British computer scientist, developed the World Wide Web in 1990. The Web has grown exponentially in the last decade and has revolutionized information dissemination.

X-ray: The German physicist Wilhelm Conrad Roentgen (1845–1923) in 1895 produced the first X-ray photograph of his wife's hand. X-rays are used in medical diagnostics and analyzing material defects in material science.

Zeppelin: The first rigid airship, or Zeppelin, was made by Ferdinand von Zeppelin (1838–1917) in 1900. Zeppelins were used for civilian and military purposes for many years before being replaced by aircraft.

INDEX

Pages mentioned in *italics* indicate pictures

PICTURE CREDITS

a-above, b-below, l-left, r-right, t-top

Kyoto National Museum
27b

Photolibrary
6b, 8, 10, 36, 46r, 46b, 54, 70b, 106

Visage Media Services/Getty Images
9b, 16, 19b, 20b, 21, 22, 28b, 30, 31, 34, 35l, 35b, 37a, 37b, 38t, 38b, 39l, 39r, 40, 41a, 41b, 42a, 42b, 43, 44, 45l, 45r, 46t, 47a, 47b, 48t, 48b, 49, 50a, 50b, 51, 52a, 52b, 53a, 53b, 55a, 55b, 56, 57l, 57r, 58a, 58l, 58r, 59, 60a, 60b, 61, 62, 63a, 63b, 64, 65a, 65b, 66t, 66b, 67, 68, 69, 70a, 71, 72, 73, 74, 75a, 75b, 76t, 76b, 77, 78a, 78b, 79, 80, 81a, 81l, 82, 83, 84a, 84b, 84, 86, 87a, 87b, 88a, 88b, 89, 90a, 90b, 91, 92, 93a, 93b, 94t, 94b, 95, 96a, 96b, 97, 98a, 98b, 99, 100, 101a, 101b, 102a, 102b, 103, 104, 105a, 105b, 107a, 107b, 108a, 108b, 109, 110a, 110b, 111

Werner Forman Archive
15, 19a, 20a
/Ashmolean Museum, Oxford 6t
/British Institute of History and Archaeology, Dar-es-Salam 7
/British Library, London 24
/British Museum, London 12, 18
/Edgar Knobloch 25
/Egyptian Museum, Cairo 13t, 17b
/E. Strouhal 17a
/Moravian Museum, Brno 9a, 11a
/Museum fur Volkerkunde, Berlin 33b
/Museum of Americas, Madrid 32b
/National Maritime Museum, Greenwich 29a
/National Museum of Anthropology, Mexico 26, 28t
/N.J. Saunders 33a
/Oriental Collection, State University Library, Leiden 23
/Private Collection, New York 27a
/Sold at Christie's, London (1996) 13b
/Spink and Son Ltd. 29b
/The Smithsonian Institution, Washington 11b